The Nuclear Power Issue:
A Guide to Who's Doing

Who's Doing What Series: 8

# The Nuclear Power Issue:
## A Guide to Who's Doing What in the U.S. and Abroad

**Edited by Kimberly J. Mueller**

**California Institute of Public Affairs**
An Affiliate of The Claremont Colleges
Claremont

Published and distributed by the California Institute of Public Affairs, P.O. Box 10, Claremont, California 91711, Telephone (714) 624-5212.

The California Institute of Public Affairs, established in 1969, is a nonprofit research and public service foundation affiliated with The Claremont Colleges. Our interests are twofold: environmental and natural resource policy (state, national, and international); and the character, problems, and future development of California. President: Thaddeus C. Trzyna. A catalog of publications is available on request.



In obtaining the data contained herein, the editors have used information supplied to them by the organizations listed and from other sources purporting to be accurate. However, neither the editors nor the publisher can be held responsible for inaccuracies or omissions which may exist in this book.

Represented outside North America by George Prior Associated Publishers, Ltd., 37-41 Bedford Row, London WC1R 4JH.



**Library of Congress Cataloging in Publication Data**

Mueller, Kimberly J
The nuclear power issue.

(Who's doing what series ; 8)
Includes indexes.
1. Atomic power--Directories. I. Title.
TK9012.M84 333.79'24'025 79-52430
ISBN 0-912102-44-6 AACR1

# Contents

# Introduction

The nuclear power issue is polarized into extreme positions: pro and anti. Each side draws leading scientists, professional societies, and voluntary citizens' organizations to its defense. On one hand, nuclear power is promoted as an important energy source for the future, when world oil supplies will be short. On the other hand, it is denounced as a source of harmful environmental radiation and unmanageable radioactive waste, as threatening democracy in that it brings greater centralization of control over energy supplies, and as encouraging nuclear weapons proliferation and nuclear terrorism through distribution of nuclear materials and technology throughout the world.

The issue is far from resolved. Despite active citizen protest, most developed countries are engaged in nuclear power programs. Most of Europe — especially France, West Germany, and the United Kingdom — pursues a nuclear path. The United States, though less single-minded, has maintained the nuclear option, and the Reagan Administration declares a strong commitment to nuclear power. Only a few developed countries have decided at present not to develop nuclear energy; these include Denmark, Australia, and New Zealand. Less-developed countries such as Brazil, Israel, South Korea, Libya, and Pakistan look to those with nuclear technology as models for nuclear development and sources of nuclear materials.

This book is designed to help people find sources of information and assistance about the nucler power issue. It includes organizations representing the whole spectrum of opinion on the issue, as well as neutral scientific bodies. We hope it will foster better communication among the many groups working on the problems associated with nuclear power, and that it will encourage and facilitate study of those problems.

The book is divided into seven major parts. *Chapters 1* and *2*, respectively, cover federal and state governmental agencies in the U.S. *Chapter 3* describes U.S. citizens', professional, research, and educational organizations concerned with nuclear power. *Chapter 4* covers nuclear power plants in the U.S., providing information (including the name and address of the responsible utility) for all facilities operating, under construction, on order, or shut down. *Chapter 5* lists international organizations, both intergovernmental and nongovernmental. *Chapter 6* is a country-by-country survey of nuclear power outside the U.S.; it includes listings of concerned governmental and nongovernmental organizations, as well as nuclear power plants, for each country. *Indexes* of organizations and acronyms and initialisms begin on page 101.

Most of the information in this book was provided directly by the organizations listed. Letters or questionnaires were sent to several hundred groups. We appreciate the help of those who responded, often in some detail. Special thanks are due to Michael Paparian, for his help in compiling information about nuclear power in California, and to Nancy Matsumoto, of our own staff, for editorial support. The information on power plants is based on documentation provided by the Atomic Industrial Forum, the U.S. Nuclear Regulatory Commission, and the International Atomic Energy Agency.

A directory such as this will never be complete. New agencies and groups emerge; others fade away as funding dries up, studies are completed, or objectives are met. Addresses are always changing. We plan to issue revisions of *The Nuclear Power Issue* from time to time. Users of this volume are urged to contribute to the next edition by sending additions and corrections to: California Institute of Public Affairs, P.O. Box 10, Claremont, California 91711.

Note: The symbol "+" before the name of an organization indicates that it reports to the organization listed above it. The symbol "+ +" means that the group is subsidiary to the "+" group above.

# Chapter 1.
# United States Government

LEGISLATIVE BRANCH:

General Accounting Office (GAO)
441 G St., N.W.
Washington, D.C. 20548 202-275-6202

An independent nonpolitical agency with information-gathering and research responsibilities. Broad authority to assist Congress (both individual members and committees) in fulfillment of its duties; makes recommendations for more effective government operations; may initiate inquiries.

In 1979, GAO issued 11 reports on NRC activities, including safeguards, emergency preparedness, radiation control, nuclear plant inspection, and nuclear materials diversion. A GAO report of March 30, 1979, recommended that a 10-mi. emergency planning zone be established around all plants, and called for better planning in areas of possible radiation hazard. The NRC is required to respond to GAO reports, to the Government Operations Committees of the House and the Senate, within 60 days.

GAO recently conducted a 5-yr. evaluation of the NRC. Its recommendations were consistent with those of the President's Commission (see below) and of a Special Inquiry Group established by the NRC itself, in calling for improved overall managerial functions of the NRC to increase reactor safety.

EXECUTIVE BRANCH:

President's Commission on the Accident at Three Mile Island
2100 M St., N.W.
Washington, D.C. 20037

Twelve-member citizens' commission created in April 1979 by Executive Order 12130 to investigate in full the accident at the Three Mile Island (TMI) Nuclear Station Unit 2 near Harrisburg, Pennsylvania (March 28, 1979). The Commission, acting on the President's mandate, looked for causes in the sequence of events, both human actions and machine operations, during the accident. It developed estimates of radiation released during the accident, investigated radiation protection to workers under normal operating conditions, and to the public if more serious circumstances were to develop; examined the organization, procedures and practices of the utility managing the TMI facility; assessed emergency plans existing at the time of the accident, state and local government response to the accident, and made recommendations to improve plans in case of future, possibly more serious accidents; and analyzed coverage of the accident to determine whether errors in news reports were due to official miscommunication or journalistic mistakes, which media sources most influenced the people directly involved in or affected by the accident, and whether media coverage exaggerated the seriousness of the accident. The Commission also investigated the NRC in terms of its structure, functions, and effectiveness in dealing with the accident.

The Commission concluded, on the basis of its investigation, that "fundamental changes will be necessary in the organization, procedures, and practice -- and above all -- in the attitudes of the Nuclear Regulatory Commission and, to the extent that the institutions we investigated are typical, of the nuclear industry." The Commission recommended a moderation of belief in "technical fixes" to solve the problem of nuclear safety. It called for a concerted effort to train nuclear facility personnel to a better technical understanding of their jobs and to a greater sense of responsibility for job performance; and it suggested that a preoccupation in the NRC with complex regulations be overcome by a deep concern for safety going beyond simple enforcement of set safety standards.

The NRC's response to the Commission's report is included in the NRC 1979 Annual Report. The Commission's report, "The Need for Change: The Legacy of TMI," submitted Oct. 30, 1979, is available from the Office of the Press Secretary to the President, The White House, Washington, D.C. 20500.

Council on Environmental Quality (CEQ)
722 Jackson Place, N.W.
Washington, D.C. 20006 202-383-1415

Provides an ongoing assessment of U.S. energy research and development from an environmental standpoint, as well as a continuing analysis of changes or trends in the environment. Concerns include the environmental aspects of nuclear power development, particularly disposal of nuclear waste and proposals for offshore nuclear generating stations. Pub: Environmental Quality, annual report on the state of the environment and measures being taken to protect it, with much material on energy; other publications.

Department of Defense Agencies

Defense Advanced Research Projects Agency (DARPA)
1400 Wilson Blvd.
Arlington, Virginia 22209 202-694-3007

Responsible for managing "high-risk, high-payoff" basic and applied research as designated by the Secretary of Defense. Performs nuclear monitoring research.

Defense Civil Preparedness Agency (DCPA)
The Pentagon
Washington, D.C. 20301 202-697-4484

Charged with providing an effective and viable National Civil Defense Program. Coordinates and provides direction to federal, state, and local governments planning in various programs, including Nuclear Civil Protection and radiological fallout warfare programs. DCPA grants radiological instruments for detecting radiation fallout from nuclear detonation to state and local governments.

Defense Nuclear Agency (DNA)
Headquarters
Washington, D.C. 20305 202-325-7046

Descended from the Armed Forces Special Weapons Project (est. by the Atomic Energy Act of 1946; became the Defense Atomic Support Agency in 1959; renamed the DNA in 1971). Central Defense agency for developing and testing nuclear weapons in coordination with the Department of Energy. Performs research and testing (inc. field experiments) of nuclear weapons effects. Manages nuclear weapons stockpile, and is responsible for security of theater nuclear forces. Advises Joint Chiefs of Staff on all nuclear matters. Also, maintains availability of Johnston Atoll in the Pacific Ocean as a test site. Is in process of restoring Eniwetok Atoll from a nuclear test site to a liveable environment so that native peoples may return to the island. DNA Field Command is located in Albuquerque, New Mexico. The Armed Forces Radiobiology Research Institute is in Bethesda, Maryland.

Department of Energy (DOE)
Washington, D.C. 20545 202-376-4000

Created by the Department of Energy Organization Act of 1977 to consolidate federal energy functions into one Cabinet-level Department. DOE assumed operating and development responsibilities in the nuclear field formerly handled by the Energy Research and Development Administration (ERDA, est. 1974 to assume functions of the old Atomic Energy Commission) and several other federal agencies.

DOE's responsibilities are to research, develop, and demonstrate energy technology; to regulate energy production, allocation and use; and to oversee the nation's nuclear weapons program. Nuclear fission programs include: supply of uranium concentrates and other raw materials; production and management of fissionable materials; development of nuclear reactors for civilian use, including safety research; and the development of the first liquid metal fast breeder reactor (LMFBR). DOE owns the Fast Flux Test Facility (FFTF), an important LMFBR test laboratory located 10 mi. north of Richland, Washington. (FFTF is not subject to NRC licensing regulations, but the NRC did review the facility in 1978 under and interagency agreement.)

DOE develops and applies systems to detect and prevent loss or diversion of nuclear materials; sponsors training in the interest of developing nuclear energy uses; and conducts studies mandated by Congress (e.g., a study begun in 1979 to review the siting of, and designate responsibility for, the West Valley, New York reprocessing plant). Until 1980, when the federal budget eliminated funds for research on gas-cooled reactors, DOE supported the Gas-Cooled Reactor Associates (GCRA), a group of utilities founded in 1978 to develop advanced commercial high-temperature gas-cooled reactors (HTGRs); and worked with the Helium Breeder Associates (HBA), an organization of utilities, and the General Atomic Company to develop and demonstrate gas-cooled

Department of Energy

fast breeder reactors (GCFRs).

DOE played a major role in monitoring radiation during the accident at Three Mile Island in March 1979. On other occasions, it supplies the NRC with technical specialists from national laboratories to assist in reviews in nuclear power plant applicants. DOE has been asked by the NRC to take the lead in assessing cost-impact of design improvements in nuclear facilities, to facilitate the transfer of new models to commercial application. Also, DOE is working to identify all sites where radioactive material was stored or processed in the early years of the nuclear age. For information on this project, contact Dr. William E. Mott, DOE, Mail Station E-201, Washington, D.C. 20545.

DOE supports a major research and engineering effort to develop nuclear fusion. Two basic approaches are being pursued: (1) confinement of fusion fuel by magnetic fields; and (2) inertial confinement of fusion, which has military as well as civilian applications. The goal of the magnetic fusion energy program is to build and operate by 1998 a demonstration power plant that will produce several hundred megawatts of electricity. The program involves close cooperation with the USSR, Japan, and the European Atomic Community (EURATOM).

DOE also supports U.S. policy on international nuclear nonproliferation and the international fuel cycle, and coordinates international energy programs. It conducts the Nonproliferation Alternative Systems Assessment Program (NASAP), for which it considers reactor types and fuel cycles and provides technical data and input to the International Fuel Cycle Evaluation (see p. 62). This function is monitored and assisted by the NRC. As outlined in the Nuclear Nonproliferation Act of 1978 (NNPA), DOE consults with the NRC on nuclear export activities, cooperation agreements, foreign distribution of nuclear materials, retransfer of U.S.-supplied nuclear materials. DOE also handles requests to reprocess irradiated U.S.-supplied nuclear fuel.

DOE programs are distributed among a number of Assistant Secretaries:

Assistant Secretary for Energy Technology. Responsible for nuclear research, development and technology demonstration. Programs include: water-cooled breeder reactors, technology development and special projects, space applications, nuclear energy assessments, light water reactor technology, advanced isotope separation technology, breeder reactor facilities, and naval reactor development. Supports nuclear fuel cycle waste management for commercial reactors, including facilities for a National Terminal Waste Storage Program, and for remedial action at 21 inactive milltailing sites and one former ore processing site. Working to develop a Waste Isolation Pilot Plant (WIPP).

Assistant Secretary for Environment. Responsible for light-water reactor safety facilities.

Assistant Secretary for Resource Applications. Responsible for uranium enrichment and research on the uranium enrichment process; also for uranium resource assessment, and voluntary and incentive programs to increase domestic supplies of uranium.

Assistant Secretary for Defense Programs. Directs national nuclear weapons research, development, testing, production, and surveillance; manages a safeguards program for special nuclear materials; and analyzes and coordinates international activities related to nuclear technology and materials.

DOE policy development and program coordination are carried out mainly at the Washington, D.C. headquarters and at nearby Germantown, Maryland. Operations are performed largely by industrial concerns and public and private institutions under contracts administered by the Department's Operations Offices:

+Albuquerque Operations Office, DOE
P.O. Box 5400
Albuquerque, New Mexico 87115 — 505-264-7231

+Chicago Operations Office, DOE
9800 S. Cass Ave.
Argonne, Illinois 60439 — 312-972-2001

+Idaho Operations Office, DOE
550 Second St.
Idaho Falls, Idaho 83401 — 208-526-7322

+Nevada Operations Office, DOE
P.O. Box 14100
Las Vegas, Nevada 89114 — 702-734-3201

+Oak Ridge Operations Office, DOE
P.O. Box E
Oak Ridge, Tennessee 37830 — 615-576-4444

+Richland Operations Office, DOE
P.O. Box 550
Richland, Washington 99352 — 509-942-7395

+San Francisco Operations Office, DOE
1333 Broadway
Oakland, California 94612 — 415-273-4237

+Savannah River Operations Office, DOE
P.O. Box A
Aiken, South Carolina 29801 — 803-824-6331

Nuclear research is performed by the DOE's national laboratories, including:

+Argonne National Laboratory
9700 S. Cass Ave.
Argonne, Illinois 60439 — 312-972-5555

Major responsibility for Liquid Metal Fast Breeder Reactor research, safety analyses, tests; manages DOE's breeder reactor components development; broad interest in fusion (magnetic and inertial).

+Brookhaven National Laboratory
Long Island, New York 11973 516-345-3335

Physics radiation research; techniques for measuring environmental pollution; major mission in fusion.

+Fermi National Acceleration Laboratory
P.O. Box 500
Batavia, Illinois 60510 312-840-3000, x3211

Elementary particle physics; technology utilization.

+Hanford Engineering Development Laboratory
P.O. Box 1970
Richland, Washington 99352 509-942-3915

Breeder reactor technology development; fast reactor safety; Fast Flux Test Facilities activities; waste management; fusion; design, development and construction projects.

+Idaho National Engineering Laboratory (INEL)
Idaho Falls, Idaho 83401 208-526-1476

Main reactor safety testing center; site of Power Burst Facility (PBF) for testing pressurized water reactors (PWR) in off-normal conditions; reactor fuel reprocessing; breeder reactor research; waste management; power plant demonstrations; the world's first high-level radioactive waste solidification process; contaminated equipment maintenance.

+Lawrence Berkeley Laboratory (LBL)
University of California
Berkeley, California 94720 415-843-2470, x5111

Principle laboratory for heavy ion research; the Bevalac heavy ion accelerator is used to explore new vistas of nuclear research; fusion research and development; radiation biology; advanced isotope separation.

+Lawrence Livermore Laboratory (LLL)
University of California
P.O. Box 808
Livermore, California 94520 415-422-7401

Conducting a nuclear waste isolation experiment whereby spent reactor fuel elements are emplaced deep in a granite formation at the Nevada test site; results are then monitored and evaluated.

+Los Alamos Scientific Laboratory
P.O. Box 1663
Los Alamos, New Mexico 87545 505-667-5061

Conducts uranium surveys for DOE's National Uranium Resource Evaluation (NURE); uranium isotope separation; models reactor accident results for the NRC; performs nuclear diversion detection research (some for the IAEA); tritium safety studies; research and development of fast reactor fuel elements.



+Notre Dame Radiation Laboratory
University of Notre Dame
Notre Dame, Indiana 45556 219-283-7502

Radiation chemistry research.

+Oak Ridge National Laboratory
P.O. Box X
Oak Ridge, Tennessee 37830 615-574-4846

Development of nuclear power as safe, economic; magnetic fusion feasibility demonstration; health and environmental effect of energy production.

+Pacific Northwest Laboratory
P.O. Box 999
Richland, Washington 99352 509-946-2201

Nuclear fuel cycle research, special concerns with safety, efficiency, environmental consequences; fusion.

+Sandia Laboratories
P.O. Box 969
Livermore, California 94550 505-264-7261

Nuclear research; fission energy; light water reactor (LWR) safety; beneficial uses of radioactive waste; nuclear waste management and deep ocean floor disposal; nuclear materials transport. Location in Albuquerque, New Mexico too.

+Savannah River Laboratory
Aiken, South Carolina 29801 803-725-6211, x3183

Prepares and reviews others' Liquid Metal Fast Breeder Reactor (LMFBR) standards, and works to develop quality assurance programs in accordance with DOE policy; tests breeder reactor parts and materials; laser fusion reactors; design work for the NRC.

Additional DOE programs and activities include the Office of Nuclear Waste Isolation (ONWI) and the National Nuclear Data Center:

+Battelle Memorial Institute
Project Management Division
505 King Ave.
Columbus, Ohio 43201 614-424-5715

Battelle operates the ONWI for DOE. It has assumed responsibility for developing long-term storage and, ultimately, isolation techniques for high-level radioactive wastes.

+National Nuclear Data Center
197D Brookhaven National Laboratory
Upton, New York 11973 516-345-2902

Gathers and manages experimental nuclear data; distributes evaluated data for fission and fusion reactors applications; coordinates data analysis and evaluation.

Department of Health, Education and Welfare (HEW)
Washington, D.C. 20852

Food and Drug Administration (FDA)
5600 Fishers Lane
Rockville, Maryland 20857 301-443-3380

FDA's Bureau of Radiological Health researches health effects of, sets safety standards for, and develops methods of controlling, radiation exposure. Works with the Environmental Protection Agency (see below) in establishing radiation criteria, standards, guidelines, and policies.

Department of State
2201 C St., N.W.
Washington, D.C. 20520 202-655-4000

Advisory agency to the President on foreign policy. Aims "to promote long-range security and well-being of the U.S." Is presently working with the Department of Energy in pursuing the possibility of a joint interim spent-fuel storage facility in the Pacific Basin to be shared by Japan.

The State Department's Bureau of Politico-Military Affairs guides internal departmental policy on issues including U.S. security and nuclear policy, and maintains liaison with the Department of Defense and other federal agencies. The Bureau of International Organizational Affairs offers policy guidance and support for U.S. participation in the International Atomic Energy Agency, United Nations, and other international organizations. The Bureau of Oceans and International Environmental and Scientific Affairs formulates and enacts U.S. policies and proposals for scientific and technological relations with other countries and international organizations, manages a range of foreign policy issues including nuclear technology, and represents the State Department in international negotiations in its area of work.

Agency for International Development (AID)
Department of State
Washington, D.C. 20523 202-632-9170

Among AID assistance programs for less-developed countries are Technical Assistance and Energy Development programs designed "to help developing countries alleviate their energy problems by increasing their production and conservation through research and development of suitable energy sources and conservation methods, collection and analysis of information concerning potential supplies of and needs for energy." AID programs also analyze the cost-effectiveness of various alternative energy sources.

Department of Transportation (DOT)
400 Seventh St., S.W.
Washington, D.C. 20590 202-426-4000

Materials Transportation Bureau
Office of Hazardous Materials Regulation
2100 Second St., S.W.
Washington, D.C. 20590 202-755-9260

The Bureau has operational responsibilities for hazardous materials transportation, including radioactive materials.

Together, DOT and NRC are conducting a surveillance program of radioactive material transported into, within, or through states agreeing to participate. In 1979, nine states were participating or had just entered the program (see pp. 20-23).

INDEPENDENT FEDERAL AGENCIES:

Environmental Protection Agency (EPA)
401 M St., S.W.
Washington, D.C. 20460 202-755-2673

Responsible for protection and enhancement of the environment; works to control pollution in several areas, including radiation.

Office of Radiation Programs, EPA
401 M St., S.W.
Washington, D.C. 20460 202-755-4894

This Office works to fulfill EPA's basic responsibility to protect the public from adverse effects of ionizing and nonionizing radiation. It conducts a national surveillance program to measure radiation in the environment; sets standards to limit radioactivity around reactor fuel-producing plants, conducts a waste-management program; provides technical assistance to states or agencies with radiation protection programs; reinforces the efforts of other federal agencies, including the NRC, against environmental pollution; and advises the President about radiation matters directly or indirectly affecting health. Guides all federal agencies in the formulation of radiation standards and in the establishment and execution of programs in cooperation with the states.

An EPA Interagency Committee on Federal Guidance for Occupational Exposures to Ionizing Radiation (also involving NRC staff) is in the process of developing occupational radiation dose limits. Also, EPA aided in evaluating the accident at Three Mile Island in March 1979.

EPA has an office in each of the 10 federal regions. It has both a western and an eastern radiation monitoring laboratory:

Environmental Monitoring and Support Laboratory, EPA
P.O. Box 15027
Las Vegas, Nevada 89114 702-736-2969

Eastern Environmental Radiation Facility, EPA
P.O. Box 61
Montgomery, Alabama 36101 205-534-7615

EPA publications include research reports on radiation, available from the Office of Public Involvement, EPA, Washington, D.C. 20460.

Federal Emergency Management Agency (FEMA)
Washington, D.C. 20472 202-653-7465

Created by Reorganizational Plan No. 3 of 1978, made effective March 31, 1979. FEMA assumed emergency responsibilities from several federal agencies, including the Department of Commerce, Department of Housing and Urban Development, and the Office of Science and Technological Policy. It is designated to prepare for and respond to major civil emergencies, including nuclear disasters. Specifically, FEMA is responsible for coordinating governmental response away from the plant site of a radiation emergency (as opposed to the NRC, which is responsible for assessing hazards at the site and recommending protective actions). FEMA coordinates its programs with those of state and local governments, private industry and voluntary organizations.

Interagency Review Group on Nuclear Waste Management (IRG)

Fourteen-member panel which conducted a year-long review of nuclear waste management. Submitted its final report to the President in March 1979. This report defined the objective which nuclear waste management should adopt as: the isolation of nuclear waste from the biosphere, with no significant threat posed to public health and safety. In general, the report urged that responsibility for establishing a waste program should not be deferred to future generations; that a broader research and development program for waste disposal, particularly geologic isolation, should begin promptly; and that public participation should be developed and strengthened for all aspects of nuclear waste management programs.

In considering repositories mined in geologic formations, the report concluded that a choice of sites with various potential host rock and geohydrological characteristics should be designated at the earliest possible date; that two or three national repositories, in different sections of the country, become operational before the year 2000; that repositories should be planned on a technically conservative basis to allow retrieval of waste for a period of time; that interim storage for spent fuel should be prepared now until disposal facilities are available (thus allowing more time to plan for permanent storage facilities); and that given the right geologic environments, reprocessing is not required to assure safe disposal of commercial spent fuel. With regards to management of uranium mill tailings, which the IRG considered a matter of great urgency, the report called for completion of NRC and EPA regulations on acceptable levels of radon emissions, siting, impacts on ground water, and ultimate disposal methods; for monitoring of abandoned sites, and remedial action at select sites; and for improved technology to handle tailings. The report also noted that decontamination and decommissioning (D and D operations) at facilities which will soon reach the end of their useful lives will require legislation to ensure adequate management.

Institutional problems of nuclear waste management, according to the IRG's report, should be subjected to a detailed national analysis. On matters of waste siting, consultation and concurrence between all parties affected should take place. Nuclear waste management agencies should strengthen mechanisms for receiving advice from expert federal agencies, professional scientific and engineering societies, review groups, scientists and engineers, the utilities, private businesspeople, public interest groups, and individual citizens.

Nuclear Regulatory Commission (NRC)
1717 H St., N.W.
Washington, D.C. 20555 301-492-7000

Main federal entity concerned with nuclear power. Created by the Energy Reorganization Act of 1974, along with the Energy Research and Development Administration (now defunct) to replace the Atomic Energy Commission (created by the Atomic Energy Act of 1946). The NRC fulfills the mandate of both the Atomic Energy and the Energy Reorganization Acts, as amended, as well as that of the Nuclear Nonproliferation Act of 1978. It is responsible for licensing and regulating nuclear materials and facilities in order to protect public health and safety, the environment (in accordance with the National Environmental Policy Act of 1969), national security, and the antitrust laws. Operational and development responsibilities are filled by the Department of Energy (see p. 10).

Broad functions of the NRC include: Licensing the possession, use, processing, transport, handling, and disposal of nuclear materials; conducting technical reviews and public hearings in the process of issuing permits and licenses to nuclear facilities; development of rules and regulations governing licensed nuclear activities; inspection of facilities and enforcement of regulations, and evaluation of overall facility operation; and performing or contracting for research in support of its various activities. For a breakdown of these functions, see the program offices of the Executive Director for Operations, below.

NRC is responsible for developing effective working relationships with the states, including assuring that adequate regulatory programs are maintained by states which exercise, by agreement with the NRC, regulatory control over certain nuclear materials activities within their boundaries (see Chapter 2, p. 19ff.). NRC also works with foreign governments regarding the regulation of nuclear energy. With the Department of Energy, it cosponsors the Nuclear Safety Information Center at Oak Ridge, Tennessee.

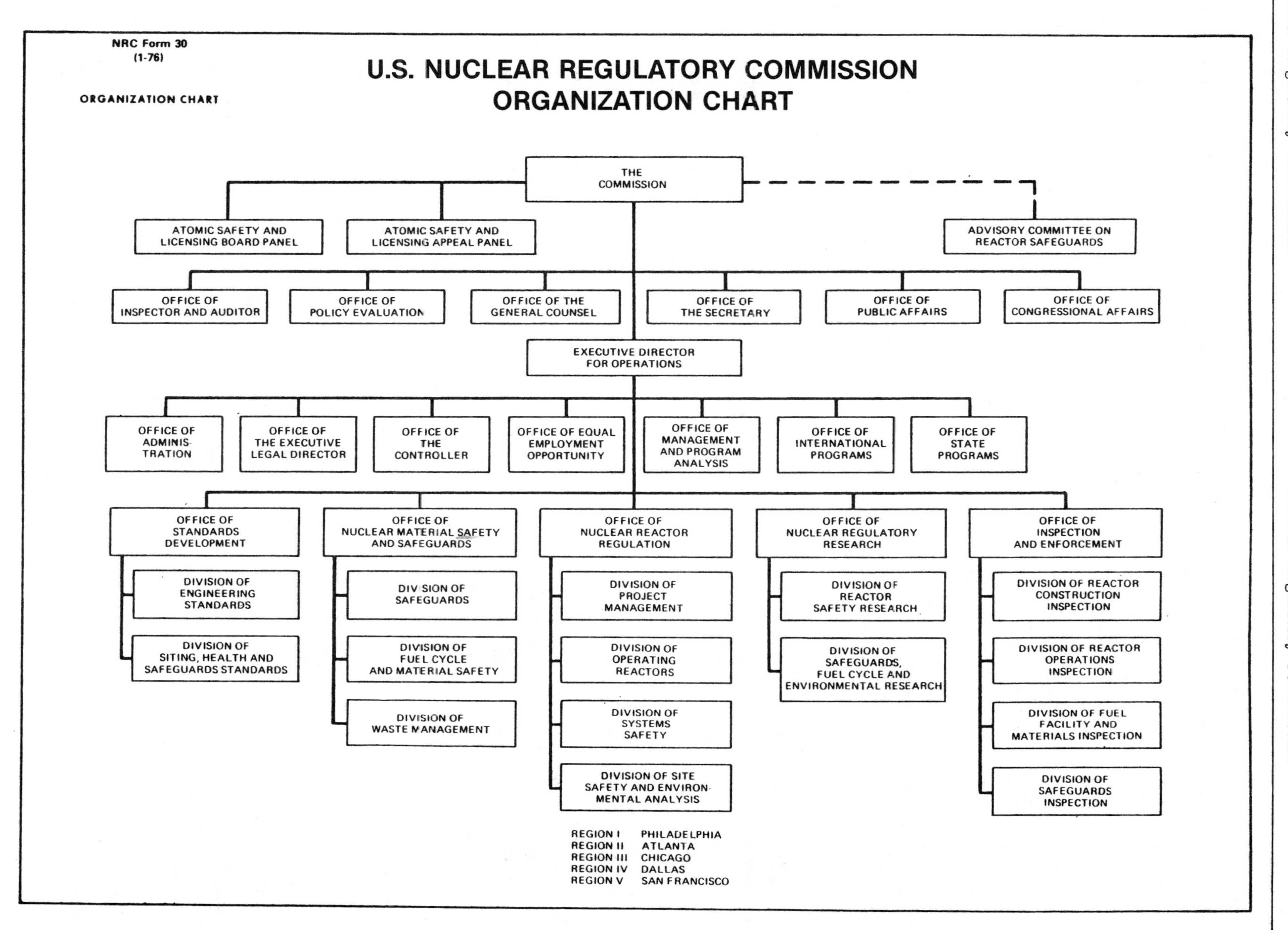
NRC Form 30
(1-76)
ORGANIZATION CHART
U.S. NUCLEAR REGULATORY COMMISSION
ORGANIZATION CHART
THE COMMISSION
ATOMIC SAFETY AND LICENSING BOARD PANEL
ATOMIC SAFETY AND LICENSING APPEAL PANEL
ADVISORY COMMITTEE ON REACTOR SAFEGUARDS
OFFICE OF INSPECTOR AND AUDITOR
OFFICE OF POLICY EVALUATION
OFFICE OF THE GENERAL COUNSEL
OFFICE OF THE SECRETARY
OFFICE OF PUBLIC AFFAIRS
OFFICE OF CONGRESSIONAL AFFAIRS
EXECUTIVE DIRECTOR FOR OPERATIONS
OFFICE OF ADMINIS-TRATION
OFFICE OF THE EXECUTIVE LEGAL DIRECTOR
OFFICE OF THE CONTROLLER
OFFICE OF EQUAL EMPLOYMENT OPPORTUNITY
OFFICE OF MANAGEMENT AND PROGRAM ANALYSIS
OFFICE OF INTERNATIONAL PROGRAMS
OFFICE OF STATE PROGRAMS
OFFICE OF STANDARDS DEVELOPMENT
DIVISION OF ENGINEERING STANDARDS
DIVISION OF SITING, HEALTH AND SAFEGUARDS STANDARDS
OFFICE OF NUCLEAR MATERIAL SAFETY AND SAFEGUARDS
DIVISION OF SAFEGUARDS
DIVISION OF FUEL CYCLE AND MATERIAL SAFETY
DIVISION OF WASTE MANAGEMENT
OFFICE OF NUCLEAR REACTOR REGULATION
DIVISION OF PROJECT MANAGEMENT
DIVISION OF OPERATING REACTORS
DIVISION OF SYSTEMS SAFETY
DIVISION OF SITE SAFETY AND ENVIRON-MENTAL ANALYSIS
OFFICE OF NUCLEAR REGULATORY RESEARCH
DIVISION OF REACTOR SAFETY RESEARCH
DIVISION OF SAFEGUARDS, FUEL CYCLE AND ENVIRONMENTAL RESEARCH
OFFICE OF INSPECTION AND ENFORCEMENT
DIVISION OF REACTOR CONSTRUCTION INSPECTION
DIVISION OF REACTOR OPERATIONS INSPECTION
DIVISION OF FUEL FACILITY AND MATERIALS INSPECTION
DIVISION OF SAFEGUARDS INSPECTION
REGION I PHILADELPHIA
REGION II ATLANTA
REGION III CHICAGO
REGION IV DALLAS
REGION V SAN FRANCISCO

The Commission is composed of five members appointed by the President, approved by the Senate. The main executive officer and official NRC spokesperson is the Chairperson.

The Executive Director for Operations oversees NRC operations and administration, and is responsible for developing policy options for NRC consideration. Program offices include:

Office of Nuclear Reactor Regulation (NRR). Conducts two-part (construction and operation) licensing process of power, test, and research reactors; monitors reactor facilities after operation commences, and has the power to decommission a facility if it does not maintain acceptable standards.

Office of Nuclear Material Safety and Safeguards (NMSS). Concerned with licensing and regulation of facilities and materials involved in processing, transport and handling of nuclear materials; reviews safety measures; recommends research, standards, and policy options where necessary to achieve its ends.

Office of Nuclear Regulatory Research. Coordinates research programs with NRC licensing and regulatory functions; research covers safety health effects and environmental impact of the nuclear fuel cycle, waste treatment and disposal, transportation of radioactive materials.

Office of Standards Development. Develops regulations and other standards for radiological health and safety, environmental protection, material and plant protection; coordinates NRC activities in national and international standards programs.

Office of Inspection and Enforcement (IE). Inspection arm of the NRC. Checks licensed nuclear facilities and materials to determine whether or not operations are conducted according to license provisions and NRC rules; inspects license applicants and facilities; investigates accidents, and reports improper actions involving nuclear materials or facilities; enforces NRC regulations and license provisions. Also manages NRC's five regional offices:

+Region I Office, NRC
631 Park Ave.
King of Prussia, Pennsylvania 19406
215-337-1150

+Region II Office, NRC
230 Peachtree St., N.W.
Atlanta, Georgia 30303 404-221-4503

+Region III Office, NRC
799 Roosevelt Rd.
Glen Ellyn, Illinois 60137 312-858-2660

+Region IV Office, NRC
611 Ryan Plaza Dr.
Arlington, Texas 76011 817-334-2841

+Region V Office, NRC
1990 N. California Blvd.
Walnut Creek, California 94596 415-486-3141

Pub: Annual Report; Report to Congress on Abnormal Occurrences, q.; Nuclear Regulatory Commission Issuances, m.; Standard Review Plan; Operating Units Status Report, m.; Regulatory Guides series; Draft and Final Environmental Statements; complete list available from Division of Technical Information and Document Control, NRC, Washington, D.C. 20555.

Independent Panels of the NRC:

Advisory Committee on Reactor Safeguards

Committee of 15 scientists and engineers, created by statute. Assesses safety aspects of existing and proposed nuclear facilities and adequacy of reactor safety standards and reports findings to the NRC. Performs other duties at Commission request.

Atomic Safety and Licensing Board Panel (ASLB)

Public participation mechanism, panel is composed of lawyers and others with technical expertise who sit on three-person licensing boards, conduct public hearings, and otherwise assist the NRC in deciding to grant, suspend, revoke, or amend NRC licenses.

Atomic Safety and Licensing Appeal Panel

Three-member appeal boards which review ASLB decisions, either in response to an appeal or on its own initiative. Decisions may be reviewed by the NRC on its initiative, or in response to petition.

Tennessee Valley Authority (TVA)
400 Commerce Ave.
Knoxville, Tennessee 37902

Woodward Bldg.
Fifteenth and H Sts., N.W.
Washington, D.C. 20444 202-566-1401

Government-owned corporation created by Congress in 1933. Manages resource development for advanced economic growth in the Tennessee Valley region. Produces electric power for municipal and cooperative systems and some federal installations. Manages the Browns Ferry nuclear plants (Units 1-3) at Decatur, Alabama. Is constructing the following nuclear facilities: Bellefonte (Units 1 and 2), Scottsboro, Alabama; Yellow Creek (Units 1 and 2), Mississippi; Sequoyah (Units 1 and 2), Daisy, Tennessee; Watts Bay (Units 1 and 2), Spring City, Tennessee; TVA Plan A (Units 1 and 2) and TVA Plant B (Units 1 and 2), Hartsville, Tennessee; Phipps Bend (Units 1 and 2), Tennessee.

U.S. Arms Control and Disarmament Agency (ACDA)
Department of State Building
Washington, D.C. 20451 202-655-4000

Develops and enacts arms control and disarmament policies to promote U.S. national security and international relations. Discusses and negotiates with other countries on arms matters, including prevention of spread of nuclear weapons to countries that do not already have them.

OTHER AGENCIES:

Nuclear matters may fall also within the jurisdiction of the following federal agencies:

The Executive Office of Science and Technology Policy.

The U.S. Army, Corps of Engineers (technical matters; supplies assistance to the NRC for its review process).

Department of the Navy, Naval Research Lab (technical assistance to the NRC for its review process).

Department of Commerce, Industry and Trade Administration (export development and trade regulation), National Bureau of Standards, and the National Technical Information Service (NTIS) (sale of federally-funded research and development reports).

Department of Housing and Urban Development, Federal Disaster Administration (FDAA).

Department of the Interior, Office of Minerals Policy, National Mine Health and Safety Academy, Bureau of Mines (mining regulation and health), and the U.S. Geological Survey (manages the National Uranium Resource Evaluation -- NURE -- program; provides technical assistance to the NRC; enforces Interior Department mining regulations).

Department of Labor, Occupational Safety and Health Administration (OSHA), Mine Safety and Health Administration.

Eximbank (finance of U.S. exports).

The National Science Foundation (basic and applied research and education in science fields).

National Aeronautic and Space Administration (NASA) (Lansat and Laser Geodynamics satellite -- Lageos -- aids miners in mineral detection; also funds radioactive waste management research).

State Government Agencies

# Chapter 2. State Government Agencies

Until recently, the states' role in nuclear power matters has varied with each state, often in proportion to the number of nuclear power plants and the level of citizen or legislative concern in a particular state. State and local governments have voluntarily, if at all, developed nuclear emergency preparedness plans. (Fourteen state plans had received NRC concurrence by the end of 1979.) Though state legislation may rule on such matters as radioactive waste handling, radioactive materials transportation, emergency response planning, and nuclear plant siting, uniformity among state laws has not been required.

The accident at Three Mile Island in March 1979 served to heighten concern among the states regarding nuclear emergencies. The federal government, in response to the accident, is taking steps to improve state emergency plans. The Federal Emergency Management Agency (see p. 14) is charged with reviewing emergency plans in all states with operating facilities; it is also to review plans in other states with facilities due to become operational soon. The FEMA reviews, when completed, can be expected to impel revisions in existing state plans. Also, NRC regional personnel have developed specific criteria for evaluation of state and local plans, which must be met before NRC concurrence is given.

Apart from the FEMA reviews, almost all state activity in nuclear matters is linked to the NRC. Governors of all fifty states have appointed a liaison officer to maintain direct communication with the NRC. State and local governments have received federal guidance and assistance through such programs as the NRC/EPA Task Force on Emergency Planning; through free training programs (in radiological emergency response operations, coordination, planning, and handling of radioactive materials in transportation accidents) offered by the NRC to qualified state and local government personnel; and through a Field Assistance Program which conducts field reviews of state radiological emergency plans.

Under the State Agreements Program (created by Sec. 274 of the Atomic Energy Act of 1954, as amended), qualified states may enter into agreement with the NRC to assume regulatory functions over byproduct and source material, and small quantities of special nuclear material. Agreement States are subject to an annual review of their radiation control programs to determine if they are compatible with public health and safety and NRC regulations. They must submit a record of "abnormal nuclear occurrences" for inclusion in a quarterly report to Congress. Those Agreement States which choose to continue regulating uranium mills and mill tailings after November 8, 1981, must adopt federal technical standards and procedures required by the Uranium Mill Tailings Radiation Control Act of 1978. Agreement States may apply for and receive technical assistance in the areas of licensing activity, health, environmental analysis, and regulatory review, and their regulatory personnel may attend NRC-sponsored technical and administrative training courses. Agreement States' radiation control program directors meet each year for discussion and to make recommendations to the NRC.

In 1979, seven states participated in a joint NRC/DOT Transportation Surveillance program; two more states entered the program late in the year. This program checks all radioactive material transported into, within, or through the borders of participating states.

Following is a list of state offices concerned with nuclear power. A number of other offices may be concerned with certain aspects of nuclear power; these may include: Departments of Public Health; Bureaus of Radiation Control, Radiological Health, or Health; Resources or Land Use agencies; Environmental Protection, Quality, or Management agencies; Public Utilities Commissions; Geological Surveys; Departments of Human Services; Labor and Industries agencies; Departments of Energy; Energy Siting Councils; Bureaus of Mines and Geology; Conservation commissions; State legislative committees; and Offices of Planning.

In the list below, Agreement States are marked (AS). States participating in the NRC/DOT Transportation Surveillance program are marked (TSP).

ALABAMA

(AS)

Office of the Governor
3734 Atlanta Hwy.
Montgomery, Alabama 36130 205-832-6960

ALASKA

Department of Commerce and Economic Development
Division of Energy and Power Development
338 Denali St.
Anchorage, Alaska 99501 907-276-0508

ARIZONA

(AS)

Atomic Energy Commission
1601 W. Jefferson, Rm. 104
Phoenix, Arizona 85007 602-255-4845

ARKANSAS

(AS)

Arkansas State Energy Office
3000 Kavanaugh Blvd.
Little Rock, Arkansas 72205 501-371-1379

CALIFORNIA

California Energy Commission (CEC)
1111 Howe Ave.
Sacramento, California 95825 916-920-6811
(Toll-free) 800-852-7516

COLORADO

(AS)

Colorado Office of Energy Conservation
1600 Downing St., Second Floor
Denver, Colorado 80218 303-839-2507

CONNECTICUT

Office of Policy and Management
Energy Division
80 Washington St.
Hartford, Connecticut 06115 203-566-3394

DELAWARE

Delaware Energy Office
114 W. Water St.
P.O. Box 1401
Dover, Delaware 19901 302-678-5644

DISTRICT OF COLUMBIA

District of Columbia Office of Planning and Development
409 District Building
1350 E St., N.W.
Washington, D.C. 20004 202-727-6365

FLORIDA

(AS)

Florida Governor's Energy Office
301 Bryant Building
Tallahassee, Florida 32301 904-488-6764

GEORGIA

(AS)

Georgia Office of Energy Resources
270 Washington St., Rm. 615
Atlanta, Georgia 30334 404-656-5176

HAWAII

Hawaii State Energy Office
Department of Planning and Economic Development
1164 Bishop St., Ste. 1515
Honolulu, Hawaii 96813 808-548-4150

IDAHO

(AS)

State Office of Energy
State House
Boise, Idaho 83720 208-334-3800

ILLINOIS

(TSP)

Commission on Atomic Energy
524 S. Second St., Rm. 415
Springfield, Illinois 62706 217-782-5057

INDIANA

Indiana Board of Health
Environmental Management Board
1330 W. Michigan
Indianapolis, Indiana 46206 317-633-8404

IOWA

Iowa Energy Policy Council
Lucas Building, Sixth Floor
Capitol Complex
Des Moines, Iowa 50319 515-281-4420

KANSAS

(AS)

Kansas Energy Office
214 W. Sixth St.
Topeka, Kansas 66603 913-296-2496

KENTUCKY

(AS) (TSP)

Kentucky Department of Energy
Office of Planning and Evaluation
Iron Works Pike
P.O. Box 11888
Lexington, Kentucky 40578 606-252-5535

LOUISIANA

(AS)

Louisiana Department of Natural Resources
Office of Conservation
Nuclear Energy Division
4545 Jamestown
Baton Rouge, Louisiana 70808 504-925-4518

MAINE

(AS)

Maine Office of Energy Resources
55 Capitol St.
Augusta, Maine 04330 207-289-3811

MARYLAND

(AS)

Maryland Department of Natural Resources
Energy Administration
Tawes State Office Building
Annapolis, Maryland 21401 301-269-2788

MASSACHUSETTS

Energy Facilities Siting Council
1 Ashburton Pl.
Boston, Massachusetts 02108 617-727-1136

MICHIGAN

(TSP)

Office of the Governor
State Capitol
Lansing, Michigan 48909 517-373-3427

MINNESOTA

Minnesota Energy Agency
980 American Center Building
150 E. Kellogg Blvd.
St. Paul, Minnesota 55101 612-296-5120

MISSISSIPPI

(AS)

Mississippi Office of Energy
Department of Natural Resources
455 N. Lamar
Jackson, Mississippi 39201 601-981-5099

MISSOURI

Missouri Department of Natural Resources
Energy Agency
1014 Madison St.
Jefferson City, Missouri 65101 314-751-4000

MONTANA

Montana Department of Natural Resources and Conservation
Energy Division
32 S. Ewing St.
Helena, Montana 59601 406-449-3940

NEBRASKA

(AS)

State Energy Office
Box 95085
301 Centennial Mall
Lincoln, Nebraska 68509 402-471-2867

NEVADA

(AS)

Public Service Commission
505 E. King St., Rm. 304
Carson City, Nevada 89710 702-885-4180

NEW HAMPSHIRE

(AS)

New Hampshire Department of Health and Welfare
Division of Public Health
Radiation Control Agency
Hazen Dr.
Concord, New Hampshire 03301 603-271-4588

NEW JERSEY

New Jersey Department of Energy
101 Commerce St.
Newark, New Jersey 07102 201-648-6293

NEW MEXICO

(AS)

New Mexico Energy and Minerals Department
113 Washington Ave.
P.O. Box 2770
Santa Fe, New Mexico 87503 505-827-2472

NEW YORK

(AS)

New York State Energy Office
Nuclear Operations
Agency Building 2, Eighth Floor
Empire State Plaza
Albany, New York 12223 518-474-7016

NORTH CAROLINA

(AS)

North Carolina Department of Human Resources
Radiation Protection Branch
1330 St. Mary's St.
Raleigh, North Carolina 27611 919-733-4283

NORTH DAKOTA

(AS)

Office of Energy Management and Conservation
1533 N. Twelfth
Bismarck, North Dakota 58501 701-224-2250

OHIO

Ohio Power Siting Commission
361 E. Broad St.
Columbus, Ohio 43215 614-466-6422

OKLAHOMA

Oklahoma Department of Energy
4400 N. Lincoln Blvd., Ste. 251
Oklahoma City, Oklahoma 73105 405-521-3441

OREGON

(AS)

Oregon Department of Energy
111 Labor and Industries Building
Salem, Oregon 97310 503-378-4128

PENNSYLVANIA

Pennsylvania Department of Commerce
419 S. Office Building
Harrisburg, Pennsylvania 17120 717-787-3003

RHODE ISLAND

(AS)

Department of Business Regulation
Public Utilities Commission
100 Orange St.
Providence, Rhode Island 02903 401-277-3500

SOUTH CAROLINA

(AS) (TSP)

Nuclear Advisory Council
2221 Devine St., Ste. 640
Columbia, South Carolina 29205 803-758-5520

SOUTH DAKOTA

South Dakota Office of Energy Policy
Capital Lake Plaza
Pierre, South Dakota 57501 605-773-3604

TENNESSEE

(AS)

Tennessee Energy Office
226 Capital Blvd., Ste. 707
Nashville, Tennessee 37219 615-741-2994

TEXAS

(AS)

Texas Energy and Natural Resources Advisory Council
Executive Office Building
411 W. Thirteenth St.
Austin, Texas 78701 512-475-5491

UTAH

Utah Energy Office
Empire Building, Ste. 101
231 E. 400 South
Salt Lake City, Utah 84111 801-533-5424

VERMONT

Vermont Energy Office
State Office Building
Montpelier, Vermont 05602 802-828-2393

VIRGINIA

Virginia State Office of Emergency and Energy Services
Energy Division
310 Turner Rd.
Richmond, Virginia 23225 804-745-3245

WASHINGTON

(AS) (TSP)

Washington State Energy Office
400 E. Union, First Floor
Olympia, Washington 98504 206-753-2417

WEST VIRGINIA

West Virginia Fuel and Energy Office
1262-1/2 Greenbrier St.
Charleston, West Virginia 25311 304-348-8860

WISCONSIN

Wisconsin Office of State Planning and Energy
1 W. Wilson St., Rm. 201
Madison, Wisconsin 53702 608-266-8234

WYOMING

Wyoming Energy Conservation Office
320 W. 25th St.
Cheyenne, Wyoming 82002 307-777-7131

# Chapter 3.
# U.S. Associations

Abalone Alliance
944 Market St., #307
San Francisco, California 94102 415-543-3910

Coalition of 64 groups, dedicated to "a permanent halt to the construction and operation of nuclear power plants in California. . . will continue until nuclear power has been completely replaced by a sane and life-affirming energy policy." Opposes nuclear energy specifically on these grounds: with conservation and rational energy use it is not needed; it is uneconomical, and leads to centralization of control over energy; nuclear power is an unacceptable health and environmental risk, and its advancement is inextricably linked to nuclear weapons proliferation.

The Alliance has served as a clearinghouse for disseminating information to, and coordinating actions of, other anti-nuclear groups around California. It has been involved in organizing major protests to nuclear power, including successful lobbying against the Sundesert nuclear plant, and, recently, the largest rally (40,000 in attendance) opposing nuclear power on the West coast. It is co-sponsoring, with the UC Labs Conversion Project (which see), statewide educational activities to "make clear the connections between nuclear power and weapons." Alliance groups appear prepared to resort to nonviolent civil disobedience to bolster their efforts.

Pub: Its About Times, m.; other publications, list available.

Alliance for Survival
712 S. Grand View St.
Los Angeles, California 90057 213-738-1041

Voluntary grassroots organization working for a non-nuclear future. Affiliated with the Abalone Alliance (see above) in northern California, and nationally with the Mobilization for Survival (see below). Organized around the slogans: Zero Nuclear Weapons, Stop the Arms Race, Ban Nuclear Power, and Meet Human Needs. First activities included opposing the San Onofre and Diablo Canyon nuclear plants. Major thrust is public education in order to arouse public concern. The Alliance has chapter contacts throughout Southern California in the areas of Los Angeles, Long Beach, San Gabriel Valley, Orange County, Riverside/Palm Springs, San Fernando Valley, Pomona Valley, the South Bay, and Thousand Oaks/Ventura County. Pub: Survival Newsletter, bi-m.

American Bar Association (ABA)
1155 E. 60th St.
Chicago, Illinois 60637 312-947-4000

National legal organization engaged in a number of public service activities in various areas, including environmental and energy law. Currently conducting a DOE-funded evaluation of state procedures for determination of power needs and regulation of power production. The evaluation includes assessment of the role of alternative types of generating technologies and of conservation, as well as the interaction between federal and state government procedures. The first phase of the project, completed April 1980, involved an in-depth analysis of the situation in Illinois and California; the second phase, in progress, involves more states as well as legislation reviews. Pub: ABA Journal, m.

American Friends Service Committee (AFSC)
Energy Education Project
Fifteenth & Chestnut
Philadelphia, Pennsylvania 19102 215-241-7000

Works to stop nuclear power and promote renewable energy alternatives. Provides workshops in non-violence training and skills for organizers. Pub: Energy for a New Society.

American Institute of Physics (AIP)
334 E. 45th St.
New York, New York 10017 212-661-9404

Federation of professional societies in physics. Pub: (contain articles of interest to scientists working in energy-related fields): Applied Physics Letters, semi-m.; Journal of Applied Physics, m.; Physics Today, m.

American Medical Association (AMA)
1776 K St., N.W.
Washington, D.C. 20006 202-857-1300

Professional medical organization. AMA's Scientific Affairs Council has issued a report concluding that nuclear power production is less dangerous than the burning of coal.

American Nuclear Energy Council (ANEC)
410 First St., S.E.
Washington, D.C. 20003 202-484-2670

Membership organization drawing on all segments of the nuclear industry which "supports development of solar, fusion and other longer range energy resources but holds that nuclear power is essential if this nation is to have adequate and dependable energy supplies during this century. . . the risks of nuclear power are minimal in relation to its public benefits." ANEC staff monitor national nuclear policy and legislation, and assess its impact on industry. ANEC maintains close links with Congress and the Administration, and provides a number of information services. Pub: Nuclear Overview, Reports to Congress, Issue Briefs, Legislative Update, Washington Wrapup, Annual Report.

American Nuclear Society (ANS)
555 N. Kensington Ave.
La Grange Park, Illinois 60525 312-352-6611

Founded 1954. ANS now comprises 12,000 individual scientists and engineers. Its main objectives are the advancement of science and engineering relating to the atomic nucleus. ANS is especially concerned with the nuclear power industry and with integrating scientific and management disciplines constituting nuclear science and technology. It encourages research, disseminates information, and holds meetings devoted to scientific and technical papers in cooperation with government agencies, educational institutions, and other organizations with similar purposes. National meetings are held each year in June and November.

ANS is organized into various divisions of members interested in a particular branch of nuclear science. Each division aims to promote the sciences and arts of its branch. Divisions include: Controlled Nuclear Fusion, Education, Environmental Sciences, Isotopes and Radiation, Materials Science and Technology, Nuclear Criticality Safety, Nuclear Fuel Cycle, Nuclear Reactor Safety, Power, Radiation Protection and Shielding, Reactor Operations, and Reactor Physics.

The ANS Standards Committee works to develop comprehensive and detailed nuclear standards, and maintains a close working relationship with the American National Standards Institute (ANSI) and other such organizations. The ANS Nuclear Power Plant Standards Committee (NUPPSCO), formed in 1978, aims to expedite the development and management of a specific group of standards. ANS Standards activities include the Information Center on Nuclear Standards (ICONS), a membership service, which provides up-to-date nuclear standards information on the various national, foreign, and international bodies involved in developing such standards.

In cooperation with the Smithsonian Science Information Exchange (SSIE), ANS maintains the Nuclear Research Information Center (NRIC), which provides prepublication descriptions of nuclear research projects. Research areas include: Nuclear Power Reactors, and Nuclear Instrumentation and Analysis in the Physical Sciences. Project descriptions are available usually two years in advance of project completion.

ANS has an office in Washington, D.C. There are local ANS sections in 34 states, as well as in Europe, Japan, and South America. There are student branches at 54 universities in the U.S. and abroad.

Pub: Nuclear News, m. (also 3 special issues per year); Nuclear Power and the Environment: Questions and Answers, 128-page booklet; Nuclear Science & Engineering, m. journal of fundamental research; Nuclear Technology, m., concerned with applications; Transactions, biennial, containing summaries of all papers presented at national meetings; Nuclear Standards News (an ICONS pub.); ANSPI/Public Information Service Newsletter, m. A Publications and Services Catalog is available on request.

The American Physical Society (APS)
335 E. 45th St.
New Yorkk New York 10017 212-682-7341

Founded 1899. Professional organization with more than 30,000 members. "Has contributed significantly to American interest and accomplishment in the world of physics in the 20th century." APS is organized into a number of specialized divisions, including a Division of Nuclear Physics and a Forum on Physics & Society. Members meet to discuss their work together. APS is a member of the American Institute of Physics (see above). Pub: Brochure.

American POWER Committee (Promote Our Wonderful Energy Resources)
395 Concord Ave.
Belmont, Massachusetts 02178 617-489-0600

National network set up and staffed by members of The John Birch Society in order "to get the federal government out of the way of the energy producers." Activities are educational and include distributing literature and setting up speaking events. Position on nuclear power is that "nuclear technology is the safest, cleanest, cheapest and most efficient way to produce large quantities of electric power. . .it is not unreasonable to claim that the nuclear method of producing large amounts of electricity is safer than the other popular methods. And when cleanliness, efficiency and cost are considered, there is simply no contest; nuclear is better." Nuclear waste disposal is held not to be a problem, nor is radioactivity in the environment. Information on the POWER Committees can be obtained also by writing: The John Birch Society, San Marino, California 91108. Pub: JBS Bulletin, m.; other literature.

Armadillo Coalition of Texas
Box 15555
Fort Worth, Texas 76119 817-921-6895

Local anti-nuclear organization; engages in some political lobbying at local and state level.

Atomic Industrial Forum, Inc. (AIF)
7101 Wisconsin Ave.
Washington, D.C. 20014 301-654-9260

National nonprofit association of organizations and individuals interested in the development and utilization of commercial nuclear energy. Its more than 600 members are located in some 20 countries and include utilities, manufacturers, architect-engineer-constructors, service companies, financial institutions, mining and milling companies, consultants, universities, labor unions, government agencies, and national laboratories. AIF activities are designed to keep the public and the nuclear industry informed on technical and economic issues.

AIF is divided into various committees, including: Breeder Policy; Domestic Safeguards; Environment; Financial Considerations; Fusion; Fuel Cycle Policy; Insurance and Indemnity; Isotope Production and Distribution; Mining and Milling; Nuclear Export Policy; Nuclear Fuel Cycle Services; Power Plant Design, Construction and Operation; Public Affairs and Information; Reactor Licensing and Safety; and Radioisotopes.

Major activities include: developing recommendations on licensing policies and procedures, including those in the area of reactor safety, and encouraging more uniform procedures for quality assurance in the nuclear industry, improvements in nuclear power plant design, construction, and operation, and addressing major national energy policy concerns such as radioactive waste handling, reprocessing of spent fuel, safeguarding of nuclear materials, financing the nuclear industry, and nuclear exports. Dealing with federal environmental requirements is also an important concern. The Forum's National Environmental Studies Project (NESP) provides its members with technical information useful in meeting the licensing requirements of the NRC.

Pub: INFO, m. newsletter for the news media and the nuclear industry; NESP, m.; Nuclear Industry, m.; Annual Report; complete list available.

Bolinas Against Nuclear Destruction
c/o Greta Goldeman
Box 361
Bolinas 94924 415-868-1120

Boston Clamshell Alliance
595 Massachusetts Ave.
Cambridge, Massachusetts 02139 617-661-6204

Direct action organization. Maintains a speakers bureau, and a labor committee that performs outreach activities to the working community. Organizes boycotts against utilities, banks, and other corporate entities involved in financing nuclear power. Works closely with other anti-nuclear organizations. Several groups have evolved out of the Alliance. Pub: No Nuclear News, m. clipping service; Newsletter; various pamphlets.

Business and Professional People for the Public Interest (BPI)
109 N. Dearborn St., Ste. 1300
Chicago, Illinois 60602 312-641-5570

Provides legal and research services to individual and citizen organizations in the Chicago metropolitan area. Areas of concern include energy and nuclear energy, and government surveillance. Since 1970, BPI has been active in challenging nuclear power for its potential safety hazards. Since 1972, it has worked to prevent construction of the Bailly Nuclear Generating Station on the Indiana Dunes National Lakeshore. BPI points out that the plant is located in an area of extremely dense population, and in an extremely fragile ecological zone as well. Pub: Annual Report and Magazine; newsletters, 2 to 3 times/yr.

Butte Environmental Council
708 Cherry St.
Chico, California 95926 916-891-6424

Serves as an educational clearinghouse for anti-nuclear and environmental groups. Pub: Homegrown, 10 times/yr.

California Citizen Action Group (Cal CAG)
491 65th St.
Oakland, California 94609 415-654-1797

Applies in California the experience and techniques used by Ralph Nader and his associates. Primary emphasis has been on consumer protection, but some work has been done on environmental issues such as energy conservation and nuclear power. Pub: Citizen Sun, every 6 wks.

California Council for Environmental and Economic Balance (CCEEB)
215 Market St., Ste. 930
San Francisco, California 94105 415-495-5666

Seeks to "balance and reconcile California's environmental and economic needs." Advocates reduction in regulation of energy facility siting (abolition of the state Energy Commission and passage of federal "fast-track" siting legislation); and rapid development of nuclear power and other energy supplies (liquefied natural gas, SOHIO oil) to assure California's future economic growth. Pub: The War Against The Atom: An Essay on Nuclear Energy (1978), booklet; Alternative Energy Sources: An Appraisal, book; Environment and Economy, m.; other publications, list available.

California Institute of Public Affairs (CIPA)
P.O. Box 10
Claremont, California 91711 714-624-5212

A foundation affiliated with The Claremont Colleges concerned with the character, public problems, and future development of California. Engaged in research, publishing, conference organization. Environmental and natural resource policy has been a major interest. Publications include: The Nuclear Power Issue: A Guide to Who's Doing What in the U.S. and Abroad; Energy: A Guide to Organizations and Information Resources in the United States; California Environmental Directory; California Energy Directory; World Directory of Environmental Organizations (with the Sierra Club). CIPA sponsors the Council on the Global Environment. Formerly: Center for California Public Affairs. Est. 1969.

California Medical Association (CMA)
925 L St.
Sacramento, California 95814 916-444-5532

Voted in 1980 to reverse its "go-slow" policy regarding nuclear power development. In defense of its move, cited a report by the American Medical Association's Scientific Affairs Council which says that the dangers of nuclear power production are actually less than those associated with the burning of coal. Pub: Annual Report.

California Public Interest Research Group (CalPIRG)
2490 Channing Way, Rm. 218
Berkeley, California 94704 415-642-9952

Involves students and community members in "research, public education, and advocacy projects that provide for a measure of critical thinking about society and lead to progressive social change." Nuclear power is a major interest. Pub: Newsletter, q.

Californians for Nuclear Safeguards (CNS)
944 Market St., Ste. 309
San Francisco, California 94102 415-543-8072

Supports stricter regulation of nuclear power plants in California than is provided currently by federal or state laws. Coordinated the effort to pass Proposition 15 on the June 1976 state ballot, which would have severely restricted nuclear power development in the state. The measure failed, but CNS continues to work for stronger implementation of existing laws in the field, including statutes enacted by the state Legislature in 1976 in response to Proposition 15.

California Tomorrow
681 Market St.
San Francisco, California 94105 415-391-7544

A publicly-supported nonprofit organization dedicated to promoting the protection of California's natural environment and the improvement of its man-made environment. Has published several in-depth articles on topics including: decommissioning of nuclear power plants, and energy conservation. Pub: Cry California, q.

Center for Law in the Public Interest (CLIPI)
10203 Santa Monica Blvd.
Santa Monica, California 90401 213-879-5588

Nonprofit public interest law firm representing individuals and groups on environmental, energy, and other social issues. Served as lead legal advisor and intervenor for groups opposing licensing of the Diablo Canyon nuclear power plant in San Luis Obispo. Pub: Public Interest Briefs, q.

Center for Development Policy (CDP)
Campaign on Runaway Reactors
225 Fourth St., N.E.
Washington, D.C. 20002 202-547-1656

The Center is a nonprofit research organization which monitors the flow of resources to developing nations, primarily from the U.S., and conducts independent nonpartisan research and analysis of development policies and their implementation. The Campaign on Runaway Reactors is a project of the Center's for conducting comprehensive research and analysis of policies and risks posed by the export of U.S. nuclear reactors and related materials to developing countries. Through the Campaign, the Center makes contact with concerned persons or groups in countries scheduled to receive nuclear materials (South Korea, the Philippines, Malaysia). It has also mobilized a number of church, environmental, consumer, and other public interest groups to sign a petition calling on the NRC to exercise greater responsibility in the area of nuclear safety. In 1980, CDP brought a suit against the EximBank, demanding that it make public its discussion of large nuclear export loans. Pub: Nuclear Export Monitor, irr.

Chico People for a Nuclear Free Future
708 Cherry St.
Chico, California 95926 916-345-8070

Citizens Against a Radioactive Environment (CARE)
2699 Clifton Ave.
Cincinnati, Ohio 45220 513-861-3533

Grassroots organization for safe energy. Main concern is to prevent the Zimmer Nuclear Power Station, which is near completion, from becoming operational. Activities include rallies, conferences, workshops, and interventions before the NRC. Pub: CARE Newsletter, m.

Citizens Against a Radioactive Environment
373 Ellis St., Rm. 501
San Francisco, California 94102 415-771-9882

Citizens Against Nuclear Power (CANP)
P.O. Box 6625
Chicago, Illinois 60680 312-786-9041

Opposes nuclear power in general, and takes its opposition to the NRC and to the courts. Holds conferences, provides speakers, distributes anti-nuclear literature. Sponsored a Rally for a Nuclear-Free Midwest in June 1980. Pub: No Nukes News, 8 times/yr.

Citizens Concerned About Nuclear Power (C.C.A.N.P.)
140 W. State St.
Ithaca, New York 14850 607-272-3040

Anti-nuclear organization. Up-to-date on nuclear issues at local, regional, and national level. Recently helped actively to push through bans on radioactive waste transport in the Finger Lakes region of New York State.

Citizens Energy Council (CEC)
Box 285
Allendale, New Jersey 07401 201-327-3914

"Since 1966 (CEC) has played a leading role in raising the consciousness of people everywhere about the world's prime task: Abolishing nuclear weapons and nuclear power plants." Provides persons to testify at hearings, and aids citizen lobbying efforts to halt nuclear fission, implement energy conservation and develop decentralized sources of renewable energy. Plans workshops and seminars, provides speakers to interested groups, rents films to members, and organizes local energy study groups. CEC also sponsors the Nuclear Hazards Information Center to carry its work by means of regional representatives throughout the country. Pub: Energy: News Digest of Nuclear Hazards vs. Alternative Energies, m.

Citizens' Energy Project (CEP)
1110 Sixth St., N.W., #300
Washington, D.C. 20001 202-387-8998

Nonprofit public interest research and advocacy organization, founded 1973. Opposes nuclear power; promotes alternative energy technologies. Since 1976 has coordinated an effort to document civil liberty violations associated with nuclear power development. CEP has a Local Energy Action Program (LEAP) to aid communities in planning ways to avoid or phase out reliance on nuclear power. Pub: Nuclear Power & Civil Liberties: Can We Have Both? (1979, 1980), book; Nuclear Power and Civil Liberties Factsheet, irr.; Nuclear Quotes Book (1978).

Citizens Environmental Coalition (CEC)
1 Main Plaza, #1016
Houston, Texas 77002 713-228-0037

Clearinghouse for environmental information for persons in the Greater Houston-Harris County area. Provides a forum for discussion of nuclear power from both sides, pro and con. Pub: CEC Newsleaf, m.

Citizens for a Better Environment (CBE)
59 E. Van Buren, Ste. 1600
Chicago, Illinois 60605 312-939-1984

Nonprofit research and litigation corporation. Program is "national in scope but local in orientation" and focuses on energy development and conservation, electric utility rate structuring, various types of pollution, and public participation and education. CBE has issued reports on the lack of evacuation plans for areas near nuclear power plants and the consequent risk to public health and safety. It has recommended changes in nuclear safety requirements and testified on nuclear emergency and evacuation plans; and testified before the Illinois House of Representatives that moratoria on nuclear power plant construction will not cause brownouts or otherwise affect the availability of electricity in northern Illinois. CBE staff appear at various speaking engagements -- on television and radio, at public functions and organizational meetings -- in states where CBE offices are located. Pub: CBE Environmental Review, bi-m. (one issue/yr. is the Annual Report); Reports and Comment papers, list available.

+Citizens for a Better Environment
2131 University Ave., Ste. 404
Berkeley, California 94704 415-549-3900

+Citizens for a Better Environment
2 West Mifflin
Madison, Wisconsin 53703 608-251-2804

CBE also has offices in Milwaukee, Wisconsin, and San Francisco, California.

Citizens for Adequate Energy
58 Sutter St., Ste. 545
San Francisco, California 94104 415-392-3210

Pro-energy group composed largely of executives of major business and agricultural groups as well as representatives from academic institutions. Offers "the vehicle to counter the disproportionately loud voices of those who oppose everything." Works in favor of nuclear, as well as various other energy, sources. Conducts seminars, provides speakers, encourages letter writing campaigns, and provides information to consumers on the total range of energy sources. Pub: Energy News, m.

Citizens for Alternatives to Nuclear Energy
P.O. Box 377
Palo Alto, California 94302

Local anti-nuclear, pro-solar organization. Pub: Rays 'N' Cane Newsletter, bi-m.

Citizens for Safe Energy (CSE)
1917A Sixteenth St.
Sacramento, California 95814 916-442-3635

Works to phase out nuclear energy and promotes safe, non-polluting alternative energy and conservation. Has worked on issues involving the Sacramento Municipal Utility District, including nuclear energy and conservation; testified at state hearings on evacuation plans; and has organized a conference on nuclear energy. Also runs a speakers bureau. Pub: CSE Newsletter, q.

Citizens for Tomorrow, Inc.
Route 1, Box 191
Rock Falls, Wisconsin 54764 715-672-5746

Concerned to protect the rights of rural landowners against the encroachment of industry, specifically utilities interested in constructing nuclear power plants. Preparing a book about one family threatened by the nuclear industry.

Citizens United for Responsible Energy (CURE)
3500 Kingman Blvd.
Des Moines, Iowa 50311 515-277-0253

Addresses safety and economic problems of nuclear energy, energy efficiency, alternative sources of energy, and nuclear arms. Pub: Currents, bi-m.

Clark University
Center for Technology, Environment, and Development (CENTED)
950 Main St.
Worcester, Massachusetts 01610 617-793-7283

Energy research in areas of nuclear power policy, radioactive waste management, nuclear power plant performance, citizen participation in power plant siting, and reform of the electricity decision process.

Clergy and Laity Concerned (CALC)
198 Broadway
New York, New York 10038 212-964-6730

Founded 1965 to work for the end of military action in VietNam; composed of 37 chapters and affiliates. CALC calls on "people of faith/ethical belief to become informed and to demonstrate their concern for peace and justice and for the dignity of human life." The main emphasis of its Human Security Program is a campaign for a nuclear moratorium "halting the development and production of nuclear weapons; stopping the construction of new nuclear plants and phasing out all the construction of new nuclear industries to safe and useful production while assuring job security for workers; and developing and using safe, renewable energy sources." CALC works actively to close or convert local nuclear facilities and to build support for national measures such as the nuclear moratorium amendment to SALT II and cutting funds for the MX missile. It has joined forces with the Religious Task Force of the Mobilization for Survival and the Riverside Church Disarmament Program (both of which see).

After the TMI accident in March 1979, CALC helped to mobilize a demonstration in Washington, D.C., dramatizing the "inescapable connection" between nuclear power and nuclear weapons. CALC also helped organize the "No More Nuclear Victims: Nuclear Moratorium Now" forum in Washington in October 1979, addressing the responsibilities of the Department of Energy with regards to the nuclear fuel cycle.

CALC chapters and affiliates are located throughout the nation. It has offices in San Francisco; Chicago; Decatur, Georgia; Philadelphia; Washington, D.C.; East St. Johnsbury, Vermont; complete list of offices available on request. Pub: Annual Report; Nuclear Moratorium Update; complete list available.

Coalition to Stop San Onofre
P.O. Box 33686
San Diego, California 92103

Coastsiders for a Nuclear Free Future
P.O. Box 1401
El Granada, California 94018 415-728-7406

Colorado Open Space Council (COSC)
2239 E. Colfax
Denver, Colorado 80206 303-393-0466

Volunteer coalition active in environmental affairs, working to "find reasonable ways to allow Colorado to develop and grow in a manner which will also save our natural heritage and way of life." Nuclear energy is one of COSC's concerns, as is uranium mining. To focus attention on mining issues, COSC sponsored a Mine Walk in June 1980, through mining regions in the state, with stops along the way for town meetings. In general, COSC operates as a clearinghouse and coordinating council for its members and for the general public. Pub: Conservator and Minewatch, m. newsletters.

Comanche Peak Life Force (CPLF)
2710 Woodmere Dr.
Dallas, Texas 75233 214-337-5885

"Nonviolent direct action group in the Gandhian tradition." Organized to stop the Comanche Peak nuclear power plant and all nuclear power and weapons facilities. Promotes life-sustaining social, economic and political alternatives to the nuclear option. Activities include: occupation of the Comanche Peak site on several occasions, followed by litigation; non-civil disobedience activities; Hiroshima anniversary observances; demonstrations; work on a conversion plan; shareholder action; participation in Survival Summer (1980); and a state/regional conference in October 1980. CPLF is a member of the Lone Star Alliance (see below).

+Glen Rose Branch Office, CPLF
312 Paluxy St.
Glen Rose, Texas 76043 817-897-4110

Committee for Nuclear Responsibility (CNR)
P.O. Box 11207
San Francisco, California 94101

Claims to be the only national anti-nuclear group which bases its activities on an argument from human rights: ". . .until we can, and do, base our anti-nuclear case on our human rights, we are condemned to lose. Although the technical arguments against nukes are extremely strong, the odds are against self-educated anti-nuke citizens gaining credibility against pro-nuke technical 'experts,' and the odds are also against the public knowing which experts to believe even when both sides present technical 'experts.' Furthermore, there is no gain possible from winning only technical arguments." CNR provides technical and simple educational materials to schools, libraries, other anti-nuclear groups, individuals. Special interests, besides the human rights issue, include the biological effects of ionizing radiation, nuclear pollution, the implications of the "benefits vs. risks" doctrine. CNR's Chairman is Dr. John Gofman, Professor Emeritus of Medical Physics at UC Berkeley, and co-discoverer of Uranium-233 (one of three nuclear fuels used in nuclear power plants). Dr. Gofman became an anti-nuclear activist after undertaking a study in 1963 for the Atomic Energy Commission of the effect of radiation on all forms of life. Pub: "Irrevy": An Irreverent, Illustrated View of Nuclear Power, book; fliers, reprints, articles, all irr. -- "CNR publishes whenever it has something new to say."

Committee for Peace and Survival
Alderson Hospitality House
Box 509
Alderson, West Virginia 24910 304-445-2769

Concerned to raise the issue of the "threat of destruction" posed by nuclear weapons. Opposed specifically to nuclear power generation in West Virginia. Engages in public education efforts; conducts demonstrations; encourages reflection on the issues. Pub: Judgment, pub. of Alderson Hospitality House, has occasional articles on the Committee's activities, q.

Committee to Bridge the Gap
1637 Butler Ave.
Los Angeles, California 90025 213-478-0829

Involved in educating the public about hazards of relatively unknown nuclear sites in Southern California. Has investigated the UCLA nuclear reactor, a partial core meltdown in the Simi Valley 20 years ago, and about six other cases. Promotes non-nuclear alternative energy sources. Pub: The UCLA Nuclear Reactor: Is It Safe?, report (1979); newsletter, irr.; journal, irr.

Community Energy Action Network (CEAN)
P.O. Box 33686
San Diego, California 92103 714-236-1684

Anti-nuclear, pro-alternative energy group. Organized coalition against the San Onofre nuclear power plant and has participated in hearings on nuclear and other energy issues. Provides literature, maintains an information center, provides speakers for debates and fora, and conducts rallies. Pub: CEAN News, bi-m.

Conference on Alternative State and Local Public Policies
The Energy Project
2000 Florida Ave., N.W., Fourth Floor
Washington, D.C. 20009 202-387-6030

Main concern is alternative energy sources, and developing legislation to motivate their development. The Conference's opposition to nuclear power is implicit in its work; it does follow legislation affecting the status of nuclear power. Pub: New Initiatives in Energy Legislation: A State By State Guide, 1979-1980, book; Ways & Means, bi-m. newsletter.

Consortium on International Peace and Reconciliation (CIPAR)
317 E. Fifth St., Ste. 8
Des Moines, Iowa 50309 515-244-2253

Group of nine church denominations which serves as an educational, communications and witness network for peace work. In 1979, CIPAR's priority was arms reduction; it sponsored as part of its program a symposium on "Salt II: Should It Be Ratified?" In 1980, priority work was in draft counseling. CIPAR is concerned primarily with nuclear weapons, though it recognizes a connection between weapons and nuclear power. Pub: CIPAR contributes to Dovetail, q., pub. of the Iowa Peace Network.

Cornell University
Program on Science, Technology and Society (STS)
632 Clark Hall
Ithaca, New York 14853 607-256-3810

An interdisciplinary unit within the University, the Program analyzes "social issues and public policy questions emerging nationally and internationally from scientific and technological expertise." Works on both governmental and nongovernmental studies; coordinates University courses, seminars, workshops and research, often with other departments or units within the University. Together with the Peace Studies Program, STS has analyzed decision-making for U.S. military research and development programs, and problems involved in using advanced nuclear converters and breeders. STS staff is composed of persons in the sciences, humanities, law, engineering, business, and public administration; special interests of persons on the staff include arms control, the anti-nuclear movement in Western Europe and the U.S., nuclear non-proliferation and nuclear export policy, among other things. Pub: Science, Politics, and Controversy: The Case of Civilian Nuclear Power in the U.S., 1946-1974, book (1979); The Atom Besieged: Extra-Parliamentary Dissent in France and Germany, book (1980); Nuclear Reactor Safety and the Role of the Congressman: A Content Analysis of Congressional Hearings, article reprint (1980); other books and reprints, complete list available.

Council on Economic Priorities (CEP)
84 Fifth Ave.
New York, New York 10011 212-691-8559

Nonprofit organization founded 1970 whose goal is that of "improving significantly the quality of corporate performance as it touches critical areas of our social and natural environment." CEP implements this goal by rigorous analysis of the facts, and in the process develops an organized data base for the subject under consideration. Recent CEP studies have included: "Power Plant Performance," a comparison of costs and efficiency of nuclear and coal-fired power (the data base for this study was purchased by the Federal Energy Administration); and an assessment of two nuclear power plants planned by the Long Island Lighting Company as unprofitable, which led to the postponement of plant construction for two years. CEP also investigates the areas of equal employment, consumer health safety, foreign investment, political influence, and military production. Pub: Newsletter, 8 to 10 times/yr.; Reports, 1 to 2/yr; Studies, 1 to 2/yr.

Critical Mass Energy Project (CMEP)
P.O. Box 1538
Washington, D.C. 20013 202-546-4790

The Project grew out of Ralph Nader's 1974 conference "Critical Mass," named for the notion that "a critical mass of people can make a critical difference." CMEP encourages citizen input to the nuclear energy policy-making process to make energy safe and efficient. It operates as an information source, maintains a national network of anti-nuclear activists, and publishes. Four nationwide conferences have been held since CMEP was formed. Pub: Critical Mass Energy Journal, m. newspaper, commentary and directory; Legislative Arts, q.; Citizens' Action Packets and Handbooks; other publications, complete list available.

Diablo Conversion Project
1945 Berkeley Way, Rm. 218
Berkeley, California 94704 415-968-8798 (day)

Coalition composed of the Research Group of the East Bay Anti-Nuclear Group, the Mid-Peninsula Conversion Project, and the Abalone Alliance Labor Task Force. Research on feasibility of the proposed repowering of the Diablo Canyon nuclear plant. Has requested that the State of California commission a study of the same. Suggests that the plant be converted to run on a fuel like natural gas. Pub: Repowering of the Diablo Canyon Nuclear Power Plant, preliminary report (1979).

Disarmament/Conversion Program
Friends Peace Committee
1515 Cherry St.
Philadelphia, Pennsylvania 19102 215-241-7230

Works to reverse the arms race and convert to a peaceful economy. Discussion, speakers bureau, information source.

EARS -- Environmental Action Resource Service
Box 545
La Veta, Colorado 81055 303-742-3221

Mail-order bookstore, access for anti-nuclear organizers. Items for sale include books, articles, reprints, posters and various other anti-nuclear articles. Pub: Nuclear Information Catalog No. 10; Directory of Nuclear Activists, out of print.

East Bay Anti-Nuclear Group (EBANG)
Research Group
1945 Berkeley Way, #218
Berkeley, California 94704 415-845-5769

Member of the Diablo Conversion Project (see above).

East-West Gateway Coordinating Council
Regional Forum
Pierce Building
112 N. Fourth St., Ste. 1200
St. Louis, Missouri 63102 314-781-9247

The Council is a bi-state association of local governments from three Illinois counties, four Missouri counties, and the city of St. Louis, founded in 1965 as a forum for cooperation among local government leaders in making planning decisions for the region. The Council's Regional Forum is a citizen advisory group to the Board of Directors which takes an active interest in nuclear issues as they affect regional planning and development. The Forum is in the process of trying to acquire for review and comment a copy of the state nuclear disaster plan from the Office of Disaster Preparedness in Jefferson City, the capital of Missouri. Pub: Annual Report; East-West Directions, bi-m. newsletter; Air Waves, m. bulletin; memos and other publications.

Educational Foundation for Nuclear Science, Inc.
1020-24 E. 58th St.
Chicago, Illinois 60637 312-363-5225

Nonprofit organization founded 1945 "to inform the public of the problems of living in the nuclear age, and to alert scientists to their new responsibilities in an age dominated by rapid advances in science and technology." Founding sponsors included Hans A. Bethe, A.H. Compton, Albert Einstein, J. Robert Oppenheimer, Linus Pauling, and Leo Szilard. The Foundation publishes the Bulletin of the Atomic Scientists, an "internationally respected authority" on science and public affairs issues, which reports on nuclear power, the arms race, third world development, alternative energy sources, the environment, human rights, nuclear proliferation, and science and citizen rights. The Bulletin logo is a clock, "symbol of the threat of nuclear doomsday hovering over mankind"; its hands, which have moved several times since the logo was adopted -- showing two minutes to midnight when the first H-bomb was exploded -- now read seven minutes to midnight. The Bulletin is published monthly except July and August.

Educomics
P.O. Box 40246
San Francisco, California 94140

Outlet for anti-nuclear publications such as "All-Atomic Comics," "Barefoot Gen," "The Anti-Nuclear Handbook," "Nuclear Dragons Attack," and "Atomic Horror."

Environmental Action Foundation (EAF)
724 Dupont Circle Building
1346 Connecticut Ave., N.W.
Washington, D.C. 20036 202-659-9682

Founded by the organizers of Earth Day 1970. EAF seeks "to provide the link between community concerns and complex environmental issues" having to do with resources, energy, and the economy. EAF works to develop ties with consumer, labor, minority, low income, senior citizen, and women's groups; it performs research and provides technical and organizing expertise "necessary for effective citizen involvement in environmental decision-making."

EAF's nine-year old Utility Project works specifically on developing economic arguments against nuclear power. The Solid Waste Project encourages the development of policies promoting integrated comprehensive and hazardous waste management; it has sponsored several "Waste Alert!" conferences with other organization.

Following the accident at TMI in March 1979, EAF produced an anthology of articles to serve as an introduction to nuclear power: Accidents Will Happen: The Case Against Nuclear Power. Another book, Switching Over: An Organizing Manual for Clean Energy, was due Fall 1980. Other publications include: The Power Line, m. newsletter; Exposure, bi-m. newsletter articulating toxics issues; Annual Report; Utility Action Guide, resource book; Waste & Toxic Substances: Resource Guide; Nuclear Power: The Bargain We Can't Afford, book; complete list available.

The Environmental Coalition
Box 757
Concord, New Hampshire 03301 603-895-9058

Alliance of more than 100 professional and conservation organizations working in the public interest to enhance local human and natural resources. One plank of its comprehensive platform calls for a reexamination of New Hampshire's guidelines on nuclear power, and for the enactment of a moratorium on long-term storage of nuclear wastes until safe methods of storage are proven. Pub: Newsletters, bulletins, irr.

Environmental Coalition on Nuclear Power
433 Orlando Ave.
State College, Pennsylvania 16801 814-237-3900

Environmental Defense Fund (EDF)
1525 Eighteenth St., N.W.
Washington, D.C. 20036 202-833-1484

National legal action organization working on a range of environmental concerns, including energy. Calls for an end to nuclear power, and its replacement with an integrated combination of alternatives; advocates restructuring of utility rates to reflect the true cost of energy consumption. In 1979, EDF intervened in the case of two nuclear power stations planned in Arizona (Palo Verde IV and V); it called for its own kind of economic analysis of the plans; plant construction plans were later dropped. Pub: EDF Letter, bi-m. newsletter for members; complete list available upon request.

+New York Office, EDF
475 Park Ave. South, 32nd Floor
New York, New York 10016

+Denver Office, EDF
1657 Pennsylvania Ave.
Denver, Colorado 80203

+Berkeley Office, EDF
2606 Dwight Way
Berkeley, California 94704

Environmentalists, Inc.
1339 Sinkler Rd.
Columbia, South Carolina 29206 803-782-3000

Official party of record as intervenor in the ongoing challenge of the licensing of the Barnwell, South Carolina, reprocessing plant. Maintains collection of resource materials on nuclear energy, "one of the largest maintained by any public interest group in the country." Provides documentation and research services to academicians, scientists and the general public. Pub: Nuclear Energy and the Reprocessing Experiment, booklet.

Environmental Policy Center (EPC)
317 Pennsylvania Ave., S.E.
Washington, D.C. 20003 202-547-6500

Founded 1972. Registered lobby organization with "the largest team of public interest lobbyists working on energy and water resource policies in Washington." EPC has three lobbyists in the nuclear policy area. Specializes in representation of various broad-based citizen coalitions before Congress and the Executive branch. EPC promotes energy conservation, renewable energy resources, and regionalized energy production systems to meet regional energy needs. It searches for solutions to problems of nuclear waste, transportation and decommissioning; works to eliminate occupational and public health exposure to low-level radiation; and lobbies for full insurance protection to the public in the event of nuclear accidents.

Environmental Policy Institute (EPI)
317 Pennsylvania Ave., S.E.
Washington, D.C. 20003 202-544-2600

Founded 1974. "National information center on energy and water resource issues." Provides information to citizens throughout the country so that they may "more effectively influence our national priorities." EPI also works to influence the Executive Branch, and to see that laws passed are implemented by the Executive as intended by Congress.

EPI promotes maximum conservation of resources. It has eight Projects, one for each major energy resource area. The Nuclear Waste Project coordinates citizen involvement to enforce federal uranium mill tailings disposal legislation; informs and assists public citizens who may wish to participate in governmental hearings and rulemakings; pushes for strong regulation of uranium mills, and monitors the Mill Tailings Radiation Control Act of 1978. The Radiation Health Information Project encourages reductions in ionizing radiation exposures, seeks funding for human health effects research, and develops policy options for stronger federal and state occupational and public health radiation protection standards.

EPI staff served as technical experts in the Karen Silkwood lawsuit brought against Kerr-McGee Corp., which, if the court fine is upheld, would mean that nuclear facilities are liable for off-site contaminations of their workers.

EPI works closely with the Environmental Policy Center (see above). Pub: Studies and books, list available.

Fellowship of Reconciliation (FOR)
Box 271
Nyack, New York 10960 914-358-4601

Religious pacifist group founded 1915. Opposes nuclear weapons proliferation, supports non-violencé and reconciliation. 50 local groups. Pub: Fellowship, m.; others, list available.

Firelands Coalition for Safe Energy (FCSE)
P.O. Box 135
Berlin Heights, Ohio 44084 419-588-2156

Educational effort directed against nuclear power, in favor of soft energy development. Members participated in the Call for a Non-Nuclear Future rally held in Washington, D.C., in April 1980. Pub: Newsletter, bi-m.

Folks United to Thwart Unsafe Radiation Emission (FUTURE)
P.O. Box 2625
Denver, Colorado 80201

Grassroots organization. Locus for opposition to uranium mining interests in Colorado. Extensive resource materials. Pub: Newsletter.

The Ford Foundation
320 E. 43rd St.
New York, New York 10017 212-573-5000

Private, nonprofit institution est. 1936, to serve the public welfare. Main function is giving funds to other institutions and organizations for work "that shows promise of helping to solve problems of national and international importance." Has supported energy studies, including the Energy Policy Project which issued a final report, "A Time to Choose: America's Energy Future, in 1974; and a commission to study nuclear power which issued in 1976 its final report, Nuclear Power Issues and Choices. In 1979, the Foundation funded a study of energy policy published as "Energy: The Next Twenty Years." This study recommended among other things that premature adoption of specific energy technologies be avoided, and that nuclear power not be abandoned but that its safety be improved. The Foundation continues to aid the search for "coherent" national energy policies. Among its projects for 1980, also, was a grant to the London School of Economics and Political Science for a conference on arms control, nuclear strategy and detente in Europe.

Pub: (in addition to those reports noted above): A Debate on "A Time to Choose," book (1977); Energy and U.S. Foreign Policy, book; Ford Foundation Letter, 6 times/yr.; Current Interests of the Ford Foundation, booklet; other books and reports, complete list available.

Foreign Policy and Research Institute (FPRI)
Science Center
3508 Market St.
Philadelphia, Pennsylvania 19104 215-382-2054

Examines international trends and issues in American foreign policy. Studies nuclear proliferation, among other things. Pub: ORBIS, A Journal of World Affairs, q. (Summer 1978 issue: "The Many Faces of Nuclear Policy").

Friends of the Bridge, Inc.
Box 232
Hager City, Wisconsin 54014

Nonprofit organization with the purpose of providing community education about energy and environmental issues, and preserving rural values. Pub: The Bridge, bi-m. newsletter which reports on local issues related to nuclear power.

Greenpeace Foundation of America
Fort Mason, Building E
San Francisco, California 94123 415-474-6767

U.S. branch of an international organization founded in Vancouver, British Columbia, in 1970. Concerned with nuclear weapons testing and nuclear energy. Works to save the great whales and harp seals from extinction.

Group Opposed to Nuclear Energy (GONE)
300 S. Tenth St.
San Jose, California 95112 408-297-2299

Anti-nuclear group. Pub: GONE Newsletter, bi-m.

Group United Against Radiation Dangers (GUARD)
144 Ave. de la Paz
San Clemente, California 92672

Opposes the San Onofre nuclear power plant.

Hartford Clamshell Alliance
P.O. Box 6346
Hartford, Connecticut 06106 203-236-2830

Educational and legal work; occasional organizing for demonstrations. Pub: Newsletter, 10 times/yr.

Headwaters Alliance
P.O. Box 7942
Missoula, Montana 59807 406-543-7022

Loose alliance of grassroots organizations throughout Montana. Goal is a nuclear-free earth; works with other groups to raise public awareness of the dangers of nuclear power. Has co-sponsored ballot initiatives: one declaring a Nuclear Free Zone in Missoula County, and one requiring very strict safety standards for any nuclear facility in the state. Other activities include demonstrations (including excursions out of state), "counter-recruiting" of high school students to the anti-nuclear anti-militaristic cause, research on nuclear materials transportation. Members are available to do special presentations for other groups; resources include audio-visual materials and movies. Pub: The Paper Sac, pub. in conjunction with the Student Action Center at the University of Montana, m.

Institute for Local Self-Reliance
1717 Eighteenth St., N.W.
Washington, D.C. 20009 202-232-4108

Private, nonprofit research and technical assistance organization founded 1974. Analyzes local city and neighborhood economies, evaluates new technologies suitable for small-scale production, and promotes democratic decision-making by instructing local residents to plan their own community and economic development; also responds to requests for information on appropriate technology, local economic development, and other aspects of local self-reliance. Current projects include a blueprint for an energy self-reliant city. Implicit in the Institute's program is an opposition to nuclear power. Pub: Self-Reliance ance, bi-m. magazine; Cities, Energy and Self-Reliance, pamphlet; How to Research Your Local Utility, booklet; other materials, list available.

Institute for Policy Studies (IPS)
1901 Que St., N.W.
Washington, D.C. 20009 202-234-9382

Center for research, education, and "social invention." Sponsors "critical examination of the assumptions and policies which define American posture on domestic and international issues, and offers alternative strategies and visions." Study concentrated in areas of energy and national security. Pub: The Counterforce Syndrome: A Guide to U.S. Nuclear Weapons and Strategic Doctrine, book (1979); The Day Before Doomsday: An Anatomy of the Nuclear Arms Race, book (1977); Journal of Social Reconstruction, q.

Institute of Nuclear Materials Management (INMM)
11704 Bowman Green Dr.
Reston, Virginia 22090 703-471-7880

Members work in government, industry, and academic institutions, and are concerned about advancing all aspects of nuclear materials management. Pub: INMM Journal, q.

Interfaith Center on Corporate Responsibility (ICCR)
475 Riverside Dr., Rm. 566
New York, New York 10115 212-870-2293

Membership organization (14 Protestant denominations, 150 Catholic communities) sponsored by the National Council of Churches. Concerned about the social impact of corporations and the application of social criteria to investments. Maintains an Energy Program which calls through stockholder resolutions for the redirection of corporate investments, away from nuclear power, and into energy conservation and renewable energy.

In 1980, ICCR members filed resolutions protesting nuclear planning development and construction with several utilities, including: Consolidated Edison (Indian Point reactors), Southern California Edison (San Onofre and Palo Verde units), and General Electric (nuclear waste storage policies). Resolutions also have been filed with UNC Resources with regards to its uranium milling process, and with Mobil and Exxon Companies with regards to their non-petroleum acquisitions, lobbying and advertising practices. Also, ICCR members initiate or participate in public hearings, testimony before government agencies, research, litigation, boycotts, publishing, open letter-writing. Pub: The Corporate Examiner, m.; a guide to companies devoted to alternative energy production and energy conservation.

Investor Responsibility Research Center (IRRC)
1522 K St., N.W., Ste. 730
Washington, D.C. 20005 202-833-3727

Founded 1972, by a group of universities and foundations "to provide impartial, concise, timely information on the social and environmental questions raised in shareholder resolutions proposed to major corporations." Nuclear power is currently a dominant issue monitored by IRRC; resolutions calling for discontinuation of investment in nuclear power are receiving from five to ten per cent support at shareholders' meetings. IRRC, in addition to its coverage of resolutions, holds occasional conferences on issues of investor responsibility and corporate accountability. Pub: The Nuclear Power Debate: Issues and Choices, book; Proxy-Issue Reports, as appropriate; News for Investors, 11 times/yr.; Special Issues Reports; Index to publications.

Kansans Against Nuclear Energy (KANE)
2921 Central Park, #4
Topeka, Kansas 66611

Kansas State University
Engineering Experiment Station
Seaton Hall
Manhattan, Kansas 66506 913-532-5844

Energy research: nuclear reactor studies, radiation shielding.

Keystone Alliance
3700 Chestnut St.
Philadelphia, Pennsylvania 19104 215-387-5254

Grassroots coalition formed in 1978 with about 25 local chapters in southeast Pennsylvania. Opposed to nuclear power; working for "a safe, democratic energy policy." Major effort has been directed against the proposed Limerick nuclear plant. Mode of operation includes: demonstrations and civil disobedience actions; public education through fairs, speaker programs, media presentations, publication; lobbying and testifying before licensing agencies; cooperative action with other groups opposed to nuclear power. Pub: Radioactivist Reporter, 6 times/yr.

Laguna Beach Alliance for Survival
30832 Driftwood Dr.
South Laguna Beach, California 92677 714-499-3190

Focuses on opposition to the San Onofre nuclear power plant. Also promotes alternative energy. Has provided educational programs and speakers to schools and community groups. Pub: SURVIVOR Newsletter, m.

League Against Nuclear Dangers (LAND)
Route 1
Rudolph, Wisconsin 54475 715-423-7996

Educational organization concerned about nuclear power plant proliferation and the dangers associated therewith. Conducts research into low-level radiation in the food chain, exposure of workers to radioactivity, emergency evacuation planning, and cancer statistics for the state of Wisconsin. Meets regularly, testifies at public hearings, also circulates a Clean Energy Petition. Pub: Newsletter, irr.; article reprints.

+Land Educational Associates Foundation (LEAF)
3368 Oak Ave.
Stevens Point, Wisconsin 54481

LAND's publishing arm. Pub: Methodologies for the Study of Low-Level Radiation in the Midwest, book; Nuclear Waste: The Time Bomb in Our Bones, booklet; Honicker Vs. Hendrie: A Lawsuit to End Atomic Power, collection of studies; other materials, complete list available.

League of Women Voters (LWV)
Education Fund
1730 M St., N.W.
Washington, D.C. 20036 202-296-1770

The general purpose of the League is to promote active citizen participation in government (by both men and women). It does this through voter education as well as action on issues the League has studied.

The Education Fund oversees a national education program on nuclear issues to provide citizens with "the objective information necessary to rationally evaluate the nuclear option and to participate effectively in these decisions." Holds conferences, plans outreach efforts, publishes. Pub: A Nuclear Waste Primer, booklet; other publications, complete list available.

Lehigh University
Energy Research Center
Packard Lab #19
Bethlehem, Pennsylvania 18015 215-816-4090

Energy research: nuclear pyysics, nuclear fuel and waste disposal, social aspects of energy use, and environmental impact of power plant thermal discharges.

Live Without Trident
P.O. Box 12007
Seattle, Washington 98102 206-324-1489

Non-violent direct action group whose goal is nuclear disarmament "with a special focus on the deadly, and costly Trident submarine program." Organizes canvassing, educational activities, mass non-violent civil disobedience actions, and rallies. On August 9, 1980, held "Survival Saturday" at Waterfront Park in Seattle in commemoration of the bombing of Nagasaki; staged a 15-min. "die-in" to simulate "what would happen if we were to ever to use one of our 30,000 nuclear bombs." Has taken court action against first-strike weapons systems production on the grounds that such production violates international law. Pub: Ballast, m.

Lone Star Alliance
4305 Avenue H
Austin, Texas 78751

Anti-nuclear coalition.

Louisiana State University
Nuclear Science Center
Baton Rouge, Louisiana 70803 504-388-2163

Research in areas of nuclear fuel handling facilities safety programs, spent fuel containers design, radiation transport code development.

Louisiana Tech University
Nuclear Center
W. Arizona St.
Ruston, Louisiana 71272 317-257-2603

Researches suitability of salt domes for nuclear waste storage.

The Manhattan Project
Box 962
Seabrook, New Hampshire 03874 603-436-5414

Direct action campaign aimed at exposing "the countless abuses forced upon the American people, by the same institutions and the same system that pushes nuclear power and weapons; to focus on nukes as just one symptom of an economic and energy system based upon corporate profit at the expense of human need; and to open discussion on the alternative directions for our society that we might choose to take." First major action was the Wall Street Action, October 28, 1979, including an anti-nuclear demonstration at the World Trade Center and blockading of entrances to the Wall Street Stock Exchange -- "the power behind the nuclear industry." Pub: Up Against the Wall Street Journal, handbook prepared in anticipation of the Wall Street Action.

Massachusetts Institute of Technology
Energy Laboratory
77 Massachusetts Ave., Rm. E19-439
Cambridge, Massachusetts 02139 617-253-3400

Est. 1972, to undertake research in a wide range of energy technologies and related economic and policy issues. The Laboratory's Nuclear Division works to advance reactor designs and technological capability needed to build them, develops models for safety and efficiency assessment, and devises methods for analyzing light water reactors and nuclear fuel cycles for the risk of nuclear weapons proliferation. Pub: e-lab, bi-m. newsletter.

Mid-Peninsula Conversion Project
867 W. Dana, Ste. 203
Mountain View, California 94041 415-968-8798

Seeks alternative to military-oriented industry, primarily through promoting solar and renewable energy employment opportunities; conducted a study of alternatives in 1978. Member of the Diablo Conversion Project (which see). Pub: Plowshare Press, bi-m.

Mobilization for Survival (MFS)
National Office
1213 Race St.
Philadelphia, Pennsylvania 19107

Nationwide coalition dedicated to reordering our national priorities. Goals are: Zero Nuclear Weapons, End the Arms Race, Ban Nuclear Power, Meet Human Needs. Opposes nuclear power as the source of dangerous radioactivity in the environment, as a producer of unmanageable nuclear waste, as the cause for destruction of Native American lands on which uranium is mined, as an encouragement to nuclear arms proliferation, and on the basis of the accident at TMI. Various groups distribute literature, provide speakers and slide shows, work in community organizing, sponsor events in the name of the anti-nuclear movement, and organize anti-nuclear demonstrations. Pub: Mobilizer, 10 times/yr.

+New York City MFS
135 W. Fourth St.
New York, New York 10012 212-673-1808

Special concern with the U.S. Department of Transportation's plans to move radioactive waste through the streets of New York City. Pub: Newsletter, bi-m.

Mountain People for Nuclear Free Life
260 Desear Way
Felton, California 95018

The Musicians United for Safe Energy Foundation, Inc.
(The MUSE Foundation, Inc.)
72 Fifth Ave.
New York, New York 10011 212-691-5422

Public charity "organized by musicians and energy activists to provide funding for qualified groups working to stop nuclear power." Especially interested in helping local "grassroots" efforts. Sponsored benefit concerts in September 1979, the proceeds from which support Foundation grants. Plans to engage in a public education/media campaign counteracting the nuclear power industry's promotions; also hopes to hold workshops to train activists in the use of media techniques.

Napa Valley Energy Alliance
2119 Brown St., Ste. 4
Napa, California 94558 707-226-1294

Affiliated with the Abalone Alliance (which see) and People United Against Rancho Seco. Raises funds for anti-nuclear activities and provides information on energy and environmental issues through speakers and distribution of films and literature. Pub: Sunspot, m.

National Academy of Sciences
Committee on Nuclear and Alternative Energy Systems (CONAES)
2101 Constitution Ave.
Washington, D.C. 20418 202-393-8100

CONAES was established in 1975 by the old Energy Research and Development Administration (ERDA) to complete a comprehensive energy study by 1977. Substantive issues have been resolved by the Committee, but the study has still not been released due to the tedious process of review, perfection of the document, and final approval.

National Association of Atomic Veterans (NAAV)
1109 Franklin St.
Burlington, Iowa 52601 319-753-6112

Assists war veterans and widows of veterans in applying for compensation from the Veterans' Administration for disabilities resulting from radiation exposure during military service. Co-sponsored the Citizen's Radiation Conference in Washington, D.C. (April 1980) along with Physicians for Social Responsibility, Clergy and Laity Concerned, the National Council of Churches, SANE, and the Environmental Policy Center (all of which see). NAAV has joined servicemen in filing suits against the VA. Pub: Atomic Veterans' Newsletter, q.

National Campaign for Radioactive Waste Safety

See: Southwest Research and Information Center.

National Committee for Radiation Victims
317 Pennsylvania Ave., S.E.
Washington, D.C. 20003 202-543-0222

Originated as the National Citizens' Hearings for Radiation Victims, held in Washington April 11-14, 1980, at which atomic veterans, widows of veterans, uranium mine workers, downwind residents, and medically irradiated persons testified before a Citizens' Commission on Radiation Policies. The Committee continues a program "to serve the needs and concerns of Americans affected by exposure to human-made ionizing radiation," has a public information service on the effects of ionizing radiation, and coordinates and encourages local action on radiation health and safety issues. Pub: Report of the Commission Panel of the Citizens' Hearings for Radiation Victims; National Committee for Radiation Victims' Newsline, m.

National Consumer Law Center, Inc.
Energy Program
11 Beacon St., Ste. 824
Boston, Massachusetts 02108 617-523-8010

Federally-funded office; represents legal interests of low-income energy consumers. Policy is to intervene "only in those matters which are consistent with its carefully chosen goals." Hopes "to achieve indirectly energy reforms for its constituency." Provides assistance and representation for consumers before administrative agencies in cases of nuclear plant siting, and is active in federal and state legislative activities, including nuclear licensing. Pub: Energy Advocates Newsletter, approx. 1/m.

National Council for Environmental Balance, Inc. (NCEB)
P.O. Box 7732
Louisville, Kentucky 40207 502-896-8731

Organization of academic scientists and engineers dedicated to "protecting Nature's environmental balance while keeping a vigorous society with adequate energy and comfortable living standards." Favors continuation of nuclear power programs. Activities include research, a speakers bureau, providing information to the public and the media, publishing. Pub: ENVIRACTION, approx. bi-m.; Nuclear Proliferation: How to Blunder into Promoting It, booklet; The Health Hazards of Not Going Nuclear, book; complete list available.

National Council of the Churches of Christ (NCCC)
Energy Project
475 Riverside Dr., Rm. 572
New York, New York 10115 212-870-2385

NCCC's Governing Board has indicated that, in the interests of an ecologically just society, it would support a national energy policy which does not rely on nuclear fission as an energy source, bans commercial processing and use of plutonium as a fuel in the U.S., and works to cultivate such a ban internationally. In October 1980, NCCC and four other religious groups organized a Responsible Energy Sabbath Weekend to mobilize the U.S. religious community to involvement in advocating responsible energy development and use. The Sabbath Weekend concept was founded in the belief that "before it is anything else, energy is a religious issue." Pub: Pamphlets.

The National Council on Radiation Protection and Measurements (NCRP)
7910 Woodmont Ave., Ste. 1016
Washington, D.C. 20014 301-657-2652

Chartered by Congress in 1964; succeeded the unincorporated National Committee on Radiation Protection and Measurements. The Council involves scientific experts working on a volunteer basis "to collect, analyze, and disseminate information and recommendations about radiation protection and measurement," and "to stimulate the exchange of ideas and promote cooperation among organizations concerned with radiation problems." NCRP plays an advisory role on both a national and an international level. It is organized in more than 50 specialized committees which review information collected in a given area. NCRP projects include: assessment of population exposure to radiation, measurement of radiation in the environment, waste disposal. Pub: Catalog of all publications available on request.

National League of Cities (NLC)
Energy, Environmental Quality and Natural Resources Committee
1620 Eye St., N.W.
Washington, D.C. 20006 202-293-7580

Founded 1924 as the American Municipal Association; name changed 1964. Aims to strengthen local government and encourage effective local leadership; "the cities' chief advocate before the Congress, the executive branch, the federal agencies and, more recently, the federal courts." Activities include: training in governance and decision-making; providing information on and evaluation of federal policy programs, legislation and regulations; monitoring federal activities and reporting on them weekly. NLC is concerned with the limits to growth implied by "the inherently finite nature of the earth's resources." The Energy, Environmental Quality and Natural Resources Committee -- one of five NLC committees -- works to draft NLC policies, some addressing the nuclear issue, which are included in the National Municipal Policy. Pub: Nation's Cities Weekly; articles, papers, handbooks; complete list available.

National Taxpayers Union (NTU)
325 Pennsylvania Ave., S.E.
Washington, D.C. 20003 202-543-1300

Public interest organization which seeks to lower taxes and cut waste in government spending. NTU also aims to point out "the economic reasons why nuclear power is neither viable nor cost effective from the standpoint of the taxpayer."

National Urban League (NUL)
500 E. 62nd St.
New York, New York 10021 212-644-6500

NUL's concern with housing, jobs and quality of life, and its growing realization of the interrelation of these issues with energy and environmental issues, led it in 1979 to establish an Energy and Urban Environment Division, charged to develop new initiatives, "serving as an advocate on behalf of the poor. . . and increasing awareness throughout the Movement of the importance of energy and the environment to the well-being of black America." The Division is concerned with nuclear, among other energy sources. Pub: Annual Report.

Natural Resources Defense Council (NRDC)
1725 I St., N.W., Ste. 600
Washington, D.C. 20006 202-223-8210

Founded 1970. Nonprofit membership organization dedicated to resource conservation and improving the environment. Active in the area of nuclear power since 1971, when it came out against the federal government's breeder reactor program; has initiated many legal actions since then. Opposed licensing of the Clinch River breeder reactor on the basis of safety problems; called for hearings on fuel reprocessing; worked for a Nuclear Non-Proliferation Act (such as the one passed in 1978).

NRDC supports the right of state and local governments to establish regulations more strict than the federal regulations existing. It also researches and testifies on the problem of radioactive waste disposal, works for legislation allowing more public participation in the nuclear licensing process, and for better protective standards for workers in nuclear plants. Recently, NRDC has become more involved than previously at the administrative level of federal government. It is also involved internationally.

Pub: Amicus Journal, q.; Summary of Legal Actions and Related Activities, As of Oct. 1, 1977, book.

+New York Office, NRDC
122 E. 42nd St.
New York, New York 10017

+Western Office, NRDC
25 Kearny St.
San Francisco, California 94108 415-421-6561

Natural Resources Guild (NRG)
833 Oakridge Ave.
North Attelboro, Massachusetts 02760 617-699-2435

Devoted to harmonizing relations between environmentalists and economists through negotiation; opposed to violent protest. Particular concern with energy and nuclear energy. Pub: Annual Report.

Nevada County People for a Nuclear Free Future
P.O. Box 457
Nevada City, California 95959 916-272-7802

Anti-nuclear, pro-alternative energy group. Conducts educational programs as well as training for non-violent action. Pub: Newsletter, bi-m.

New Directions
305 Massachusetts Ave., N.E.
Washington, D.C. 20002 202-547-6600

A citizens' lobby headed by the former administrator of the Agency for International Development. Works for nuclear arms control as a means of mediating U.S.-U.S.S.R. tensions, and of setting a role model for other countries. In a statement issued April 1980, New Directions indicated that the spread of nuclear capability to non-nuclear nations is the greatest danger posed by nuclear weapons in the 1980s. It also recommended that the U.S. give complete ratification to the SALT II treaty, complete negotiations on a comprehensive ban on nuclear weapons testing, work for a treaty to halt production of fissionable material for weapons testing, promote an effective international system of plutonium and U-235 management to ensure nuclear fuel supplies but control availability of fissionable materials, and insist on Pakistan's compliance with nuclear safeguards despite the problems posed by the situation in Afghanistan. Pub: New Directions Newsletter, bi-m.; The Truth About Arms Control, pamphlet; other publications, list available.

New England Coalition on Nuclear Pollution, Inc. (NECNP)
Box 637
Brattleboro, Vermont 05301 802-257-0336

Founded 1971. First action was to challenge, through legal channels, siting of the Vermont Yankee Nuclear Power Station. NECNP also acted as legal intervenor in the licensing of the Seabrook, New Hampshire, plant, and has opposed other New England plants. "Imperfect and unbalanced as (the federal licensing process) is, it provides a legal way to force consideration of vital issues which would otherwise be largely ignored." The Coalition also works to educate the public through media presentation, a mobile energy van, a speakers bureau. Pub: On the Watch, m. newsletter; pamphlets; films; complete list available.

The New England Energy Congress
14 Whitfield Rd.
Somerville, Massachusetts 02144

Bi-partisan organization composed of all 25 members of the U.S. House of Representatives from the six New England states. Sponsored by the New England Congressional Caucus and Tufts University; funded by grants from the Economic Development Administration, U.S. Department of Commerce, and the Department of Energy's Office of the Environment. In its final report, released May 1979, the Congress reviewed the energy situation in New England (which is heavily dependent on oil), and made recommendations for the future. It urged that the nuclear power option for electricity (then accounting for 9.6 per cent of the electricity supply mix) be maintained, but that all environmental, social and economic costs be applied to the final decision on its continuation. Nuclear licensing, said the Congress, should be reformed both to shorten the process and to assure quality initial reviews of proposed nuclear plants. It also said that equal licensing procedures should be applied to heavy water and other types of reactors. The Congress urged that a national waste disposal facility for high-level nuclear waste be operational by 1988-1992 (as recommended by the President's Interagency Review Group), and if that deadline is not met, that future construction of nuclear plants be halted. A final recommendation was that uraniferous and thorium-bearing granites be explored as a potential fuel source. Pub: Final Report of the New England Energy Congress: A Blueprint for Energy Action.

The New York Academy of Sciences
Conference Department
2 E. 63rd St.
New York, New York 10021 212-838-0230

Membership organization for those involved in the advancement of science. After the accident at TMI in March 1979, the Academy sponsored a conference, "The TMI Nuclear Accident: Lessons and Implications." The conference proceedings were published as one of the Academy's Annals. Pub: Annals, approx. 25 times/yr.

New Mexico Peace Conversion Project
2405 Meadow Rd., S.W.
Albuquerque, New Mexico 87105 505-873-0299

Opposes nuclear energy on both humanistic and economic principles. Particular concern for New Mexico, with its plethora of government-funded nuclear activities. Maintains an outreach service which provides speakers and workshops, films, literature and study groups. Pub: Swords into Plowshares, q. newsletter.

Northern Sun Alliance
1519 Franklin Ave., East
Minneapolis, Minnesota 55408 612-874-1450

Nuclear Energy Women (NEW)
7101 Wisconsin Ave.
Washington, D.C. 20014 301-654-9260

Organization of women professionals in the energy industry sponsored by the Atomic Industrial Forum (which see). Aims "to promote knowledge and understanding about energy (by) providing the facts about nuclear power." Especially concerned to raise the consciousness of women about energy issues. Conducts energy education programs for women. Activities and services include a speakers bureau, energy workshops and seminars, publication and audio-visual information, energy tours, energy education exhibits, and news media services. Pub: Fact sheet.

Nuclear Transportation Project
c/o American Friends Service Committee
92 Piedmont Ave., N.E.
Atlanta, Georgia 30303 404-586-0460

National project engaged in research of the "highly-developed transportation system supporting nuclear power and weapons production." Serves as an information source and organizing catalyst for local groups. Monitors federal legislation and agency regulatory action; prepares occasional reports and analyses of federal actions and environmental impacts of transportation of nuclear materials. Advocates letter-writing and other, legal, means of reaching elected and governmental officials with the anti-nuclear message. Activities having to do with nuclear power are coordinated with the Critical Mass Energy Project (which see). Pub: Action Alert: Nuclear Waste Transportation; newsletter, irr.

Operation Turning Point (OPT)
777 United Nations Plaza
New York, New York 10017 212-490-0010

Concerned with nuclear power as it relates to nuclear weapons proliferation. Sponsors a public education program, leadership training, conferences to influence government, business and labor; also provides funding for other organizations working for disarmament. Pub: List available.

Oregon State University
Department of Nuclear Engineering
Radiation Center C100
Corvallis, Oregon 97331 503-754-2344

Review and development of boiling water reactors (BWRs); analysis of pressurized water reactor (PWR) cores; study and evaluation of fission-product after heat.

Oregon State University
School of Oceanography
Oceanography I-414
Corvallis, Oregon 97331 503-754-4763

Energy research; subsealed disposal of high-level nuclear wastes.

Oregon Student Public Interest Research Group (OSPIRG)
918 S.W. Yamhill
Portland, Oregon 97205 503-222-9641

Branch of the national Public Interest Research Group (which see). Performs research on dumping grounds for radioactive wastes; work used as the basis for the 1980 Oregon initiative concerning nuclear safeguards and a possible moratorium on new construction.

Pacific Alliance
4111 24th St.
San Francisco, California 94114 415-285-5797

Public education about safe energy. Activities include production of benefit concerts and rallies, and the sale of T-shirts. Makes grants to local groups promoting their activities.

Pacific Studies Center
867 W. Dana, Rm. 204
Mountain View, California 94041 415-969-1545

Library and information center on political and economic aspects of (nuclear) energy development and related environmental issues, particularly in the Pacific region. Pub: Pacific Research, q.; 500 Mile Island: The Philippine Nuclear Deal, study of nuclear energy development in the Philippines; other publications, list available.

Palmetto Alliance
2135-1/2 Devine St.
Columbia, South Carolina 29205

Anti-nuclear coalition.

Paumanok People's Organization
333 Terry Rd.
Smithtown, New York 11787 516-360-0045

Affiliated with the SHAD Alliance (which see). Organizes SHAD-sponsored events. Non-violent direct action. Holds local educational meetings. Pub: Fallout Forum, discontinued.

Pelican Alliance
P.O. Box 596
Point Reyes, California 94937 415-663-8483

Group of West Marin residents working to stop construction and operation of nuclear power plants and promote alternatives to nuclear power. Designed and distributed a mock-check for insert in utility bills saying "No Dollars for Nuclear Power." Pub: Pelican Paper, m.

People Against Nuclear Power (PANP)
944 Market St., Rm. 808
San Francisco, California 94102 415-781-5342

Public education activities: films, speakers, public demonstrations. Organizes small groups for participation in San Francisco. PANP activities are overseen by the Abalone Alliance (which see).

People for a Nuclear Free Future (PNFF)
515 Broadway
Santa Cruz, California 95060 408-425-1275

Group of Santa Cruz County residents working on "public education and non-violent direct action in support of nuclear disarmament and non-nuclear energy production. Pub: Newsletter, irr.

People for Energy Progress (PEP)
P.O. Box 777
Los Gatos, California 95030 408-996-8666, x777

Pro-nuclear organization "formed to advocate the development and implementation of an energy policy that permits continuing improvement of the standard of living of present and future generations." Activities have included assisting in pro-Diablo Canyon nuclear power plant rallies, participation in Department of Energy and California Energy Commission activities, distribution of "Free Speech Messages" to radio stations, sales of "Nukes Keep America Running" T-shirts and baseball caps, providing speakers for local groups and debates, and providing legislators with information. Pub: PEP-Line, bi-w.

People Generating Energy
452 Higuera St.
San Luis Obispo, California 93401 805-543-8402

Working to prevent the Pacific Gas and Electric Diablo Canyon nuclear power plant from going into operation. Pub: Newsletter, irr.

Physicians for Social Responsibility, Inc. (PSR)
P.O. Box 144
Watertown, Massachusetts 02172 617-924-3468

Certified physicians who feel a "responsibility to alert the public to the unprecedented threat to public health posed by nuclear power"; who cite three problems of nuclear power as possibly "insoluble": radioactive waste, the danger of nuclear accidents, nuclear weapons proliferation. In February 1980, PSR organized, and Harvard and Tufts Medical Schools sponsored, a symposium entitled "The Medical Consequences of Nuclear Weapons and Nuclear War," which produced an open letter from more than 700 signing physicians to leaders of the U.S. and the Soviet Union. The letter was hand-delivered to the White House and the Soviet Embassy in Washington in March. It appeared in The New York Times March 2, in the PSR April Newsletter, and has since been reprinted in the Bulletin of the Atomic Scientists.

Currently, the President of PSR is Dr. Helen Caldicott, well-known speaker against nuclear power. PSR has chapters in Old Westbury, New York; Philadelphia, Pennsylvania; Boston, Massachusetts; New Haven, Connecticut; Columbus, Ohio; Des Moines, Iowa; San Francisco, California; Palouse, Washington; and Toronto, Canada.

Pub: PSR Newsletter, q.; Medical Hazards of Radiation Packet; Health Dangers of the Nuclear Fuel Chain and Low Level Ionizing Radion -- A Bibliography and Literature Review; other brochures, books, films, videocassettes and slide shows, complete list available.

Pollution and Environmental Problems (PEP)
Box 309
Palatine, Illinois 60067 312-381-6695

Public interest citizens' organization, founded Earth Day 1969. Since 1971 has concentrated on nuclear power issues. Has lobbied representatives to oppose nuclear power, sponsored conferences, performed research, and worked to educate the public. Pub: PEP-Talk, q. newsletter.

Prairie Alliance
P.O. Box 2424, Sta. A
Champaign, Illinois 61820 217-384-4103

Public interest group concerned for the environment, opposed to nuclear power. Promotes renewable energy alternatives. Pub: Prairie Alliance Newsletter, m.

The Progressive Foundation
315 W. Gorham St.
Madison, Wisconsin 53703

Fosters public debate, discussion of issues of current and urgent concern. Especially concerned with nuclear power and nuclear arms. Pub: Time Bomb: A Nuclear Reader From The Progressive (1980).

Protect the Peninsula's Future (PPF)
P.O. Box 1677
Sequim, Washington 98382 206-683-6644

Founded in the 1970s to protest a nuclear power plant proposed for the Miller (Blyn) Peninsula between Port Townsend and Sequim. Construction of the nuclear plant has since been cancelled. PPF's present concerns are with combatting Alaskan oil pipelines. Pub: Newsletter, irr.

Public Interest Research Group (PIRG)
1346 Connecticut Ave., N.W.
Washington, D.C. 20036 202-833-3934

Public policy research arm of the Ralph Nader organization. Energy, and particularly nuclear power safety and alternate sources, has been a major interest. Pub: Pamphlets, papers; complete list of publications available.

Public Media Center
25 Scotland St.
San Francisco, California 94133 415-434-1403

Nonprofit advertising agency which helps public interest groups present their views effectively. Recent work has included asking for fairness doctrine time to counter advertisements promoting nuclear power. Pub: Newsletter, annual.

Puerto Rico Nuclear Center
Bio-Medical Building
Cabara Heights Station
San Juan, Puerto Rico 00922

Center for graduate research and advanced training; special interest in medical applications of nuclear technology.

Radiation Research Society
4720 Montgomery Lane
Bethesda, Maryland 20014 301-654-3080

Rancho Seco Defense Fund
294 Page St.
San Francisco, California 94102

Project of the Capp Foundation, to raise funds to allow presentation of the anti-nuclear case at the trial of thirteen demonstrators charged with trespassing on the Rancho Seco nuclear plant site. The Fund prepared pre-printed postcards addressed to the Governor, calling for the shutdown of Rancho Seco.

The Rand Corporation
1700 Main St.
Santa Monica, California 90406 213-393-0411

Public interest policy research organization.. The Rand Energy Policy Program currently focuses on four broad policy areas, including management of the nuclear fuel cycle. Topics of study include: the link between nuclear power generation and nuclear weapons proliferation; international arrangements for uranium enrichment; subnational issues such as terrorism, affecting nuclear power systems; international cooperation in spent fuel management; the future of government-nuclear industry relations, especially after TMI; and the way in which information about risk (technical risk assessment and public perceptions of risk) can be applied to the design of safety programs and regulatory decisions. Pub: Evaluating Nuclear Power: Voter Choice on the California Nuclear Initiative, book; Nuclear Power and the Public Safety: A Study in Regulation (1979); other publications, complete list available.

Redwood Alliance
P.O. Box 293
Arcata, California 95521 707-822-7884

Anti-nuclear group. Pub: Newsletter, m.

Rhode Islanders for Safe Power
Box 69
Wakefield, Rhode Island 02880

Anti-nuclear group; special concern for nuclear plant workers. "Why should Nuclear Workers be classified 'NOT Members of the Public'?"

Riverside Alliance for Survival
200 E. Blaine St.
Riverside, California 92507 714-748-0047

Riverside Disarmament Program
490 Riverside Dr.
New York, New York 10027 212-749-7000

Founded 1978. Program of the Riverside Church (pastor is William Sloane Coffin, anti-nuclear activist). Opposes nuclear power, especially as it encourages nuclear weapons proliferation. In October 1979, sponsored a conference, along with the Riverside Research Institution, to reverse the arms race. Pub: Disarming Notes, m.

Rocky Flats/Nuclear Weapons Facilities Project
1660 Lafayette St.
Denver, Colorado 80218 303-832-4508
or 914-358-4601

Rocky Flats is the name of a nuclear weapons plant where plutonium triggers for U.S. warheads are made. The Project is a national organization sponsored jointly by the Fellowship of Reconciliation and the American Friends Service Committee (which see). Opposed to nuclear power and dedicated to making nuclear weapons projects more visible to the public as a means of discrediting them. Sponsors demonstrations, organizes vigils at corporate meetings, works with local and national legislators. Pub: Monthly mailings.

Rocky Mountain Greenpeace
2029 E. Thirteenth Ave.
Denver, Colorado 80206 303-355-7397

Environmental and conservation interests, with hopes of stimulating "practical, intelligent actions to stem the tide of planetary destruction" hastened by nuclear power proliferation. Pub: Greenpeace Examiner, q.

Roses Against a Nuclear Environment
4400 Fair Oaks Ave.
Menlo Park, California 94025

Safe Energy Coalition (SECO)
17736 Five Points
Detroit, Michigan 48240 313-531-8943

Nonviolent nonpartisan organization composed of concerned citizens devoted to raising public awareness, encouraging mass action against nuclear power, and promoting environmentally responsible sources of energy. Works with the media; arranges speaking events, including debates with pro-nuclear parties; promotes a moratorium on nuclear power in Michigan. Has acted as intervenor in the case of the Fermi-II nuclear plant. Is investigating the possibility of organizing utility rate resistance. Pub: SECO Newsletter, q.; Resources, irr.

+Wayne State University SECO (WSU-SECO)
4866 Third St., Rm. 300
Detroit, Michigan 48202 313-577-3480

Safe Energy Communication Council (SECC)
1536 Sixteenth St., N.W.
Washington, D.C. 20036

Members represent various consumer, environmental and public interest-media groups (including Campaign for Political Rights, Critical Mass Energy Project, Center for Renewable Resources, Earth Day '80, Environmental Action, Media Access Project, National Public Interest Research Group, Solar Lobby, Nuclear Information and Resource Service) who are concerned to challenge "the well-financed public relations campaign of the nuclear industry." SECC means to communicate "accurate, reliable, factual information" to the public, and also to keep in touch with public views on the nuclear energy issue. Pub: A Citizen's Guide to Countering the Nuclear Industry's Media Blitz (January 1980).

Safe Power for Maine (SPM)
P.O. Box 774
Camden, Maine 04843 207-236-3610

"Citizens' voice" volunteer organization. Acted as intervenor in utility rate case involving Central Maine Power Company, which proposed rate hikes to recover costs of its cancelled Sears Island nuclear power project. Has petitioned the Public Utilities Commission in protest of plans to use ratepayers money to expand the spent fuel pool at the Maine Yankee plant. SPM acts as an intervenor for the public on other occasions as well. Staff members have testified before the NRC, and have served on state committees which decide on nuclear issues.

SANE: A Citizens' Organization for a Sane World
514 C St., N.E.
Washington, D.C. 20002 202-546-4868

Leads a program for conversion of (nuclear) weapons industries to peaceful, more productive civilian works. Organizes local community efforts, educates the public, and works to affect relevant legislation. Several labor unions have joined SANE in these projects. Pub: "More Jobs: Converting to a Peaceful, Productive Economy," brochure; Conversion Reading Packet.

Santa Barbara People for a Nuclear-Free Future
312-I E. Sola St.
Santa Barbara, California 93101 805-966-4565

Provides educational materials (films, video, seminars, workshops, discussion materials) on the dangers of nuclear power and weapons. Coordinates with other community organizations on related activities. Pub: People for a Nuclear-Free Future Newsletter, irr.

Santa Clarans Opposing Nuclear Energy (S.Ç.O.N.E.)
c/o San Jose Peace Center
300 S. Tenth St.
San Jose, California 95112 408-297-2299

Science for the People (SftP)
897 Main St.
Cambridge, Massachusetts 02139 617-547-0370

Loose decentralized association of about 20 organizations. Works to demystify science in order to make it responsive to people's needs. Sees a need also for a change in the existing social and political structures before the people are served properly. Activities include: workshops on nuclear power, organized study trips to China, political theater, debates, study groups and demonstrations. Pub: Science for the People, bi-m. magazine; China: Science Walks on Two Legs, book; complete list available.

Scientists and Engineers for Secure Energy, Inc. ($SE_2$)
570 Seventh Ave.
New York, New York 10018 212-840-6595

Founded 1976 following the California referendum on nuclear power to correct the misunderstanding of scientific and technological issues prevalent in the national energy debate. $SE_2$ is dedicated to "prudent advancement of the use of technology for the benefit of mankind." Its members attempt to counter the anti-nuclear "scare campaign" of several activist and doctrinaire groups. After the TMI accident, $SE_2$ operated as an information source for the press. It also has issued a report pointing out inaccuracies and fallacies in the movie "China Syndrome," and analyzed the 1979 anti-nuclear March on Washington. $SE_2$ staff have testified on behalf on nuclear power plant licensing appeals, and have called for a lift of the ban on further construction of nuclear power plants imposed after the TMI accident. Pub: Status Update, approx. 1/mo.

Scientists' Institute for Public Information (SIPI)
355 Lexington Ave.
New York, New York 10017 212-661-9110

Founded 1963 by the late Margaret Mead, Barry Commoner, and other scientists to "impartially communicate information to nonspecialists about scientific topics of concern." Works to disseminate information on scientific issues before they become emergencies, but also serves as an important source of information and interviews with specialists for journalists and other media-persons in emergencies. Received many calls during and after the accident at TMI (averages in general 75 calls a week). Not long after the TMI accident, SIPI co-sponsored a conference held at Harvard University's Kennedy School of Government to assess the impact of the accident on the public's attitude towards technology.

In October 1978 SIPI contracted with the Tennessee Valley Authority (TVA) to organize a public participation program to help decide whether spent nuclear fuel should be stored at plant sites in densified storage pools or at an away-from-reactor (AFR) storage pool. The contract expired in October 1979, at which time TVA set up its own citizen's action office to continue the program.

SIPI organizes task forces on topics of public concern, and seminars for congressmen (e.g., one in 1976 on the economics of nuclear power). It awards Student Science Journalism Fellowships, as well as the Margaret Mead Internship in Policy Related Science. Pub: Environment, 10 times/yr. (published in cooperation with the Helen Dwight Reed Educational Foundation, Washington, D.C.); SIPI Scope, bi-m. newsletter; The Economic Viability of Nuclear Energy and Nuclear Power: Economics and the Environment (1978 task force reports); other materials, list available.

The SEA Alliance (Safe Energy Alternatives)
324 Bloomfield Ave.
Montclair, New Jersey 07042 201-744-3358

Coalition of anti-nuclear, pro-safe energy activists whose purpose is to educate the public to nuclear technology's dangers. Pub: SEA News, bi-m.

+Morris County SEA Alliance
Box 271
New Vernon, New Jersey 07976 201-538-6676

Local chapter of the New Jersey SEA. Educational focus: lectures, debates, literature tables, meetings, films. Has targeted, through civil disobedience actions and a rate-withholding campaign, the utility which owns and operates the TMI nuclear power plants. Runs a special Campaign to Stop Uranium Mining, whereby action is taken against mining interests (including several major oil companies) in northern New Jersey.

Seacoast Anti-Pollution League (SAPL)
5 Market St.
Portsmouth, New Hampshire 03801 603-431-5089

Founded in 1969 to protect the coastal environment from damage resulting from various sources, particularly the Seabrook (New Hampshire) nuclear power plants. Legal intervenors in the Seabrook licensing proceedings since their onset in 1972. Helped convince the NRC, on environmental and procedural grounds, to order stay of construction in both 1977 and 1978. Continues to intervene in the Seabrook case while expanding efforts to other coastal problems. SAPL's basic program is to prevent pollution, particularly from radiation, and to preserve New Hampshire's way of life. It sponsors film and lecture series, debates, talks. Pub: SAPL News, q.

The SHAD Alliance (Sound-Hudson Against Atomic Development)
New York City Region
339 Lafayette St.
New York, New York 10012 212-475-4539

The Alliance is a regional coalition of grassroots organizations opposed to nuclear power (on grounds of its inherent dangers and economic unreasonableness) and supportive of alternative energy technologies. SHAD appeals to nonviolent direct action to raise public awareness of the nuclear issues. Activities include educational films or slide shows, discussions, leafletting, door-to-door canvassing, as well as civil disobedience actions. These latter are focussed presently at the Shoreham and Indian Point nuclear plant sites. SHAD operates in a decentralized fashion, with an annual conference as its main decision-making forum; decisions are reached by consensus or not at all. Pub: The Rising Sun, bi-m.; pamphlets, fliers.

Branches of SHAD, other than the New York City office, include:

+Long Island Region/SHAD Alliance
P.O. Box 972
Smithtown, New York 11787 516-360-0045

+Mid-Hudson Region/SHAD Alliance
Energy Action Project
S.U.B., Rm. 313
SUNY at New Paltz
New Paltz, New York 12562

+New England Thruway SHAD
642 Pelham Rd.
New Rochelle, New York 10805 914-632-2630

+Rockland/Orange Region/SHAD Alliance
57 Elysian Ave.
Nyack, New York 10960 914-358-7001

+Westchester/Putnam Region/SHAD Alliance
255 Grove St.
White Plains, New York 10601 914-682-0488

Sierra Club
Radioactive Waste Campaign
Box 64, Sta. G
Buffalo, New York 14213 716-832-9100

The Sierra Club, with national headquarters in San Francisco, is active in a wide variety of environmental concerns. The Radioactive Waste Campaign is an educational effort to increase public awareness of nuclear waste issues, thereby increasing public participation in the decision process regarding these issues. Particular concern is for the West Valley radioactive waste site, as well as other New York locations where radioactive waste is situated. The Campaign disseminates facts about radioactive waste, focussing on areas such as health hazards, worker exposure, waste disposal, available technologies and transportation. It helps local groups to organize, and provides speakers to other organizations. Pub: The Waste Paper, q.; fact sheets, technical papers, other literature, complete list available.

Smithsonian Science Information Exchange (SSIE)
1730 M St., Rm. 300
Washington, D.C. 20036 202-634-3933

Sources of information on research (most federally-funded) in progress in various scientific fields; includes research on nuclear energy, and the sociology, economics, and politics of energy development. There is a fee for services. Complete list of reports available.

Snake River Alliance
P.O. Box 1731
Boise, Idaho 83701

Society Uniting for Non-Nuclear Years (SUNNY)
P.O. Box 8
Pacific Grove, California 93950

Sonoma Alternatives for Energy
P.O. Box 452
Sonoma, California 95476 707-996-5123

Educates Sonoma area residents about the hazards of nuclear energy and alternatives to nuclear power through distribution of a monthly newsletter. Also provides training in civil disobedience. Actively involved in attempts to close Rancho Seco, the nuclear power plant operated by the Sacramento Municipal Utility District (SMUD). Pub: Safe Times, m.

So No More Atomics (SONOMA)
883 Sonoma Ave.
Santa Rosa, California 95404 707-526-7220

Committed to a "permanent halt to the construction and operation of nuclear power plants in California." Works to insure that no more money is spent on nuclear power reactors; that energy policy shifts to conservation and renewable energy; that people who lose jobs from the nuclear industry be provided training in alternative energy areas; and that the production, testing, and stockpiling of nuclear weapons be stopped. Pub: Newsletter.

Southern States Energy Board (SSEB)
1 Exchange Place, Ste. 1230
2300 Peachford Rd.
Atlanta, Georgia 30338 404-455-8841

Energy research. Maintains a Nuclear Energy Center to assess feasibility and practicality of a large number of nuclear power plants at one site. Oversees nuclear facility siting demonstration projects.

Southwest Research and Information Center (SRIC)
P.O. Box 4524
Albuquerque, New Mexico 87106 505-242-4766

Source of technical and legal information and research findings for community groups in New Mexico and elsewhere in the Southwest. Special emphases include: environmental impacts of uranium mining, nuclear waste disposal, health effects of radiation. SRIC's National Campaign for Radioactive Waste Safety is designed to oppose the federal government's Waste Isolation Pilot Plant Project (WIPP), planned for construction at a salt bed near Carlsbad, New Mexico. The Campaign conducts scientific and technical research, testifies before government committees and agencies, conducts public relations activities, disseminates information on waste management, works with other concerned groups. Pub: The Workbook, bi-m. magazine; Nuclear Waste News, bi-m. newsletter; New Mexico Uranium Inventory; complete list available.

Tehamans Against Nuclear Power
905 Jackson, No. 2
Red Bluff, California 96080 916-527-8054

Texas A&M University
Center for Tectonophysics
College Station, Texas 77843 713-845-3251

Geological studies, including isolation of nuclear waste in salt and in crystalline rocks.

Texas A&M University
Department of Nuclear Engineering
College Station, Texas 77843

Energy research. Works on improvement of nuclear reactor design procedures.

Texas A&M University
Department of Physics
College Station, Texas 77843

Research to develop nuclear theory; controlled thermonuclear fission and fusion research.

Texas Mobilization for Survival
c/o American Friends Service Committee
1022 W. Sixth St.
Austin, Texas 78703

Twin Cities Northern Sun Alliance
1519 E. Franklin Ave.
Minneapolis, Minnesota 55404 612-874-1540

Founded 1977 to block construction of nuclear plant in Tyrone, Wisconsin; initial efforts were successful. Program continues to work against nuclear power and nuclear weapons development, involving labor, church, and other sectors of the community. Devoted to the reassertion of citizens' rights over corporate and governmental authority. Concerned also for farmers' land rights and Native Americans' treaty rights. Activities include research and public education, direct non-violent action such as fasting, public demonstrations, site occupations. Pub: Northern Sun News, m.

Union for Radical Political Economics (URPE)
National Office
41 Union Square West, Rm. 901
New York, New York 10003 212-691-5722

An umbrella organization for people "devoted to the study, development, and application of radical political economics as a tool for building socialism in the United States." Activities include publication, local and regional speakers bureaus, conferences and classes, defense and advance of interests of URPE members. URPE's 1980 Summer Conference mandated that energy receive special attention from the organization. A weekend conference on energy-related issues to be held in New York City is planned for early 1981. Topics which have been suggested for discussion include: world trade in nuclear technology, siting of nuclear plants or "other disruptive or dangerous facilities," ecological and occupational health aspects of nuclear power, well-known critiques of current energy programs (Amory Lovins's work, Carter's National Energy Plan, CONAES, Ford Foundation report), the potential of the anti-nuclear movement. URPE has Regional Coordinators located in Berkeley, California, and Portland, Oregon. Members of its Steering Committee are located all around the country. Pub: URPE Newsletter, bi-m.; The Review of Radical Political Economics, q.; Dollars and Sense, m. bulletin; URPE Reading Lists; Economics Education Project series; pamphlets.

Union of Concerned Scientists (UCS)
1208 Massachusetts Ave.
Cambridge, Massachusetts 02138 617-547-5552

Founded 1969. Prominent coalition of some 70,900 engineers and other professionals. Performs advocacy work relevant to controversial issues such as nuclear power. A Scientists' Declaration on Nuclear Power, prepared in 1975 under UCS auspices, reads: ". . .it now appears imprudent to move forward with a rapidly expanding nuclear power plant construction program. The risks of doing so are altogether too great. We, therefore, urge a drastic reduction in new nuclear power plant construction starts before major progress is achieved in the required research and in resolving present controversies about safety, waste disposal, and plutonium safeguards. For similar reasons, we urge the nation to suspend its program of exporting nuclear plants to other countries pending resolution of the national security questions associated with the use by these countries of the by-product plutonium. . ." UCS addresses the problems of safety, radioactive waste disposal, and plutonium safeguards through research, public education, publishing. Among UCS's founders was Prof. Henry Kendall of MIT. Pub: The Nugget File (government information on accidents and safety defects); Looking But Not Seeing -- The Federal Nuclear Power Plant Inspection Program, report; The Risks of Nuclear Power Reactors (review of NRC safety study); others, list available.

U.S. Student Pugwash Committee and Conference
c/o History of Science
Yale University
2036 Yale Station
New Haven, Connecticut 06520 203-436-3445

Modeled after the International Pugwash Conferences on Science and World Affairs which encourage scientists in particular to heed their moral responsibilities to seek solutions to world problems and to explore the links between technology and ethical dilemmas. The first Student Pugwash Conference was held in San Diego in June 1979, to facilitate interdisciplinary examination of world problems, to allow students to consider these problems early in their professional development, to identify new questions, and to discuss timely concerns with experts. The Conference is divided into five workshops, three of which are: Energy, the Economy, and the Environment; Weapons and World Peace; Regulation of Science and Technology. Students are chosen to attend by an application process.

University of California, Berkeley
Department of Nuclear Engineering
4105 Etcheverry Hall
Berkeley, California 94720 415-645-5107

Research into nuclear reactor theory and safety; radiation detection; nuclear fuel cycles; radioactive waste management; the environmental impact of nuclear technology; fusion.

University of California, Berkeley
Earth Sciences Division
1 Cyclotron Rd.
Berkeley, California 94720    415-486-4000

Energy research, including studies of underground nuclear waste storage.

University of California, Berkeley
Materials and Molecular Research Division
1 Cyclotron Rd.
Berkeley, California 94720    415-486-4000

Nuclear physics research; experimentation with laser treatment of nuclear waste.

University of California, Los Angeles
T.H. Hicks Nuclear Energy Laboratory
405 Hilgard Ave.
Los Angeles, California 90024    213-825-2040

Study of nuclear reactors and fuels, nuclear engineering, and nuclear waste management.

University of California, San Diego
Energy Center
La Jolla, California 92093    714-452-4284

Seeks to "strengthen interdisciplinary programs of research and teaching as well as to provide graduate and post-doctoral students with added research opportunities, facilities and assistance." The Center uses grant money for most of its programs and draws on faculty from various departments. Has conducted research into the environmental effects of large nuclear farms, and fission, breeder, and fusion reactors.

University of Iowa
Free Environment
Activities Center IMU
Iowa City, Iowa 52242    319-353-3888

Clearinghouse for environmental problems, projects and inquiries. Available to both the University and the community. Main interest is cultivation of a wilderness ethic, but also involved in anti-nuclear activities, mainly through liaison with other organizations.
Pub: Free Environment Magazine, 10 times/yr.

University of Michigan
Department of Nuclear Engineering
151 Cooley Building
Ann Arbor, Michigan 48109    313-764-4262

Research programs include: evaluation and analysis of low enrichment fuel, magnetic fusion energy research (theoretical), fusion reactor engineering.

University of Michigan
Michigan Memorial-Phoenix Project
Phoenix Memorial Laboratory
North Campus
Ann Arbor, Michigan 48109    313-764-6213

Founded 1949 to promote peaceful uses of nuclear energy. Awards faculty grants for research into such uses. Operates the Ford Nuclear Reactor and the Phoenix Memorial Laboratory, both facilities for nuclear research and education. Provides research services. Staff participates in a program sponsored by the Argonne National Laboratory, to develop a low enrichment uranium reactor fuel which could be used to convert existing high enrichment plants, reducing diversion risks and safeguard costs of handling and storing fuels. Pub: Annual Report.

University of Missouri Research Reactor Facility (MURR)
Research Park
Columbia, Missouri 65211    314-882-4211

Observation of the neutron; fundamental nuclear science activities; nuclear engineering; effects of radiation.

University of Notre Dame
Department of Aerospace and Mechanical Engineering
Notre Dame, Indiana 46556    219-283-4245

Energy engineering research, including response of Liquid Metal Fast Breeder Reactors to heat and stress.

University of Tennessee
International Energy Symposia Series
Energy, Environment, and Resources Center
329 S. Stadium Hall
Knoxville, Tennessee 37916    615-974-6063

Symposia involving national and international leaders in the energy field to consider world energy productivity and production, future paths of energy production and use, and alternative energy policies. The first session was held October 1980; the second is planned for June 1981, the third for May 1982. Planned in conjunction with the 1982 World's Fair, to take place in Knoxville, with the main theme of energy. The Symposia are sponsored by the U.S. Department of Energy, the International Energy Agency, the Tennessee Valley Authority, and the University of Tennessee.

University of Virginia
Department of Nuclear Engineering and Engineering Physics
Reactor Facility
Observatory Mountain
Charlottesville, Virginia 22901    804-924-0311

Research into Liquid Metal Fast Breeder Reactor accidents, safeguards for nuclear fuel cycles, self-protection of fuel elements. Citizens' workshop program.

University of Wisconsin -- Madison
Department of Nuclear Engineering
153 Engineering Research Building
Madison, Wisconsin 53706 608-263-1648

Studies of heat removal in fast reactors after accidents, emergency core cooling in light water reactors (LWRs), nuclear fuel elements, fusion technology, and laser fusion.

University of Wisconsin -- Madison
Nuclear Reactor Lab
130 Mechanical Engineering Building
Madison, Wisconsin 53706 608-262-3922

Experimental nuclear technologies research.

University of Wisconsin -- Milwaukee
College of Engineering and Applied Sciences
P.O. Box 784
Milwaukee, Wisconsin 53211 414-963-5126

Energy research, including nuclear reactor safety.

Upper Napa Valley Energy Alliance
1513 Madrona Ave.
St. Helena, California 94574 707-963-7835

Provides information about the anti-nuclear movement. Periodically raises funds for the Mobilization for Survival and other national groups with which it is affiliated. Presents speakers, films and tapes at regularly scheduled monthly meetings. Pub: Upper Napa Valley Energy Alliance Newsletter, m.

Upstate People for Safe Energy Technology, Inc.
(UPSET, Inc.)
Box 264
Hermon, New York 13652

Ventura Safe Energy Council
P.O. Box 1966
Ventura, California 93001

Anti-nuclear group which provides information on current energy topics. Members participated in non-violent training sessions in anticipation of the possible opening of the Diablo Canyon nuclear power plant.

Victor Gruen Center for Environmental Planning

Terminated 1980. Research collection deposited at the Southern California Institute of Architecture in Santa Monica. Before termination, in June 1979, co-sponsored a conference, "The Nuclear Crisis: A Public Dialogue," to explore the nuclear energy issue.

Wall Street Action
339 Lafayette St.
New York, New York 10012 212-673-0680

Part of The Manhattan Project (which see). A direct action campaign "aimed at raising public consciousness about the way corporations and the financial community control people's lives. . . (focussing) on nukes as just one symptom of an economic and energy system based upon corporate profit at the expense of human need." The Action's focal point was a "takeover" of the New York Stock Exchange on October 29, 1979 (fiftieth anniversary of the stock market crash in 1929), preceded on October 28 by a legal and educational rally. Pub: Flier.

War Resisters League (WRL)
339 Lafayette St.
New York, New York 10012 212-228-0450

Activities stress the link between nuclear weapons and nuclear power; include demonstrations, non-violent direct action, dissemination of educational materials, organization and training. Pub: WRL News, bi-m.; WIN (co-published by WRL, available from 326 Livingston St., Brooklyn, New York 11215), bi-w.; Peace Calendar, annual.

Westchester People's Action Coalition (WESPAC)
255 Grove St.
White Plains, New York 10601 914-682-0488

Founded 1973. Coalition of organizations working for social change of one kind or another; a base for the anti-nuclear movement and training ground for non-violent direct action. WESPAC spawned the SHAD Alliance (which see) which is now a major anti-nuclear force in the state of New York. It has participated in occupations of the Seabrook nuclear facility, and organized demonstrations at Indian Point. Concerned also with disarmament. Has a library and resource center; counseling and lobbying services. Pub: Newsletter, bi-m.

Western New York Peace Center
440 Leroy Ave.
Buffalo, New York 14215 716-835-4073

Interested mainly in disarmament. Has designed a program and a book, "A World for Our Children: The Case for a Non-nuclear Future." Pub: The Peace Center Report, q.

Wilmington College
Peace Resource Center
Pyle Center, Box 1183
Wilmington, Ohio 45177 513-382-5338

Concerned with nuclear power as it relates to nuclear weapons proliferation. Maintains the Hiroshima/Nagasaki Memorial Collection which preserves, for the purposes of public education, materials relating to the atomic bomb attacks on Japan in 1945. Pub: Newsletter, q.; Annotated Bibliography of Japanese A-Bomb Literature, book; other publications, list available.

Wolf Creek Nuclear Opposition
Westphalia, Kansas 66093

Defunct.

Women of All Red Nations (WARN)
P.O. Box 2508
Rapid City, South Dakota 57701

Concerned with radioactive contamination (e.g., of water) on the Pine Ridge Indian Reservation.

Woodstock Nuclear Opponents
P.O. Box 604
Woodstock, New York 12498 914-679-6307

Presents public education programs about nuclear dangers, street theater programs, fund-raising concerts.

Worldwatch Institute
1776 Massachusetts Ave., N.W.
Washington, D.C. 20036

Independent nonprofit research organization. Publishes papers on global problems for decision makers. scholars, and the general public. Paper No. 6, published May 1976, is entitled "Nuclear Power: The Fifth Horseman," by Denis Hayes. Information on other publications available from the Institute.

Nuclear Power Plants in the U.S.

# Chapter 4. Nuclear Power Plants in the United States

On January 1, 1980, according to the Atomic Industrial Forum, Inc., the U.S. had 72 nuclear power reactors with operating licenses, with a total electrical generating capacity of 52,366 megawatts (1 megawatt = 1 million watts), or about 9.5 per cent of the nation's total electrical capacity. Ninety-one reactors had construction permits (total capacity, 100,052 megawatts); 4 reactors had limited work authorizations (4,112 megawatts), and 25 reactors were on order (29,094 megawatts).

Federal licensing of nuclear plants was suspended after the accident at Three Mile Island in March 1979. It was resumed in mid-1980. However, due to the NRC's efforts to apply careful, more stringent standards to proposed plants, as well as to public resistance to some plants -- and also to utilities' reassessment of the economic feasibility of nuclear (especially with regard to insurance liability in case of a nuclear accident) -- licensing proceeds slowly.

When all reactors under construction or on order are completed, nuclear power will account for nearly 20 per cent of the nation's electrical generating capacity. Regionally, nuclear's share in this capacity is even greater: in Wisconsin, Maryland, Arkansas and South Carolina, it represents 30 per cent of electrical generation; in Nebraska and Connecticut, about 50 per cent; in Maine, 65 per cent; and in Vermont, 79 per cent. A full 38 per cent of New England's electricity is supplied by nuclear power.

The following list includes U.S. commercial nuclear plants with operating licenses, construction permits, limited work authorizations, or on order. It is arranged alphabetically by state, and within each state by operating utilities. For each plant is given its name, location, reactor type, net power output (in megawatts), and its date of actual or planned commercial operation.

ALABAMA

Alabama Power Company
600 N. Eighteenth St.
Birmingham, Alabama 35291 205-323-5341

Operating: Joseph M. Farley 1, Houston County. Pressurized water reactor. Net power output: 829 MW. Commercial operation: December 1977.

Construction permit: Joseph M. Farley 2, Houston County. Pressurized water reactor. Net power output: 829 MW. Commercial operation: October 1980.

Tennessee Valley Authority (TVA)
Commerce Building
400 Commerce Ave.
Knoxville, Tennessee 37902 615-632-2101

Operating: Browns Ferry 1, 2, and 3, at Decatur. Boiling water reactors. Net power output: 1,065 MW each. Commercial operation: August 1974, March 1975, and June 1977, respectively.

Construction permit: Bellefonte 1 and 2, at Scottsboro. Pressurized water reactors. Net power output: 1,235 MW each. Commercial operation: September 1983 and June 1984, respectively.

ARIZONA

Arizona Public Service Company
The Salt River Project
P.O. Box 21666
Phoenix, Arizona 85036 602-271-7900

Construction permits: Palo Verde 1, 2, and 3, at Wintersburg. Pressurized water reactors. Net power output: 1,270 MW. Commercial operation: May 1983, May 1984, and May 1986, respectively.

ARKANSAS

Arkansas Power & Light Company
P.O. Box 551
Little Rock, Arkansas 72203 501-371-4000

Operating: Arkansas Nuclear One-1 and -2, at Russellville. Pressurized water reactors. Net power output: 850 MW and 912 MW, respectively. Commercial operation: December 1974 and 1980, respectively.

CALIFORNIA

Pacific Gas and Electric Company (PG & E)
77 Beale St.
San Francisco, California 94106 415-781-4211

Construction permits: Diablo Canyon 1 and 2, at Avila Beach. Pressurized water reactors. Net power output: 1,084 MW and 1,106 MW each. Commercial operation: 1981?

On order: Two units. Boiling water reactors. Net power output: 1,168 MW each. Commercial operation: no date yet established.

Shut down: Humboldt Bay, at Humboldt Bay. Boiling water reactor. Net power output: 65 MW. Closed July 1977 for seismic modification. Commercial operation: uncertain.

Sacramento Municipal Utility District (SMUD)
6201 S St.
Sacramento, California 95813 916-452-3211

Operating: Rancho Seco 1, at Clay Station. Pressurized water reactor. Net power output: 918 MW. Commercial operation: April 1975.

Southern California Edison Company
P.O. Box 800
Rosemead, California 91770 213-572-1212

Operating: San Onofre 1, at San Clemente. Pressurized water reactor. Net power output: 436 MW. Commercial operation: January 1968.

Construction permits: San Onofre 2 and 3, at San Clemente. Pressurized water reactors. Net power output: 1,100 MW each. Commercial operation: October 1981 and January 1983, respectively.

Nuclear Power Plants in the U.S.

COLORADO

Public Service Company of Colorado
550 Fifteenth St.
Denver, Colorado 80202 303-571-7511

Operating: Fort St. Vrain, at Platteville. High temperature gas-cooled reactor. Net power output: 330 MW. Commercial operation: July 1970.

CONNECTICUT

Connecticut Yankee Atomic Power Company
107 Selden St.
Berlin, Connecticut 06037 203-666-2431

Operating: Haddam Neck, at Haddam Neck. Pressurized water reactor. Net power output: 575 MW. Commercial operation: January 1968.

Northeast Nuclear Energy Company
c/o Northeast Utilities
P.O. Box 270
Hartford, Connecticut 06101

Operating: Millstone 1 and 2, at Waterford. Boiling water and pressurized water reactor, respectively. Net power output: 660 MW and 870 MW, respectively. Commercial operation: March 1971 and December 1975, respectively.

Construction permit: Millstone 3, at Waterford. Pressurized water reactor. Net power output: 1,150 MW. Commercial operation: May 1986.

FLORIDA

Florida Power Corporation
3201 34th St., South
St. Petersburg, Florida 33733 813-866-5151

Operating: Crystal River 3, at Red Level. Pressurized water reactor. Net power output: 825 MW. Commercial operation: March 1977.

Florida Power & Light Company
9250 W. Flagler St.
Miami, Florida 33174 305-552-3552

Operating: Turkey Point 3 and 4, at Turkey Point. Pressurized water reactors. Net power output: 693 MW each. Commercial operation: December 1972 and September 1973, respectively.

St. Lucie 1, in St. Lucie County. Pressurized water reactor. Net power output: 802 MW. Commercial operation: December 1976.

Construction permit: St. Lucie 2, in St. Lucie County. Pressurized water reactor. Net power output: 802 MW. Commercial operation: May 1983.

Nuclear Power Plants in the U.S.

## NUCLEAR POWER PLANTS IN THE UNITED STATES

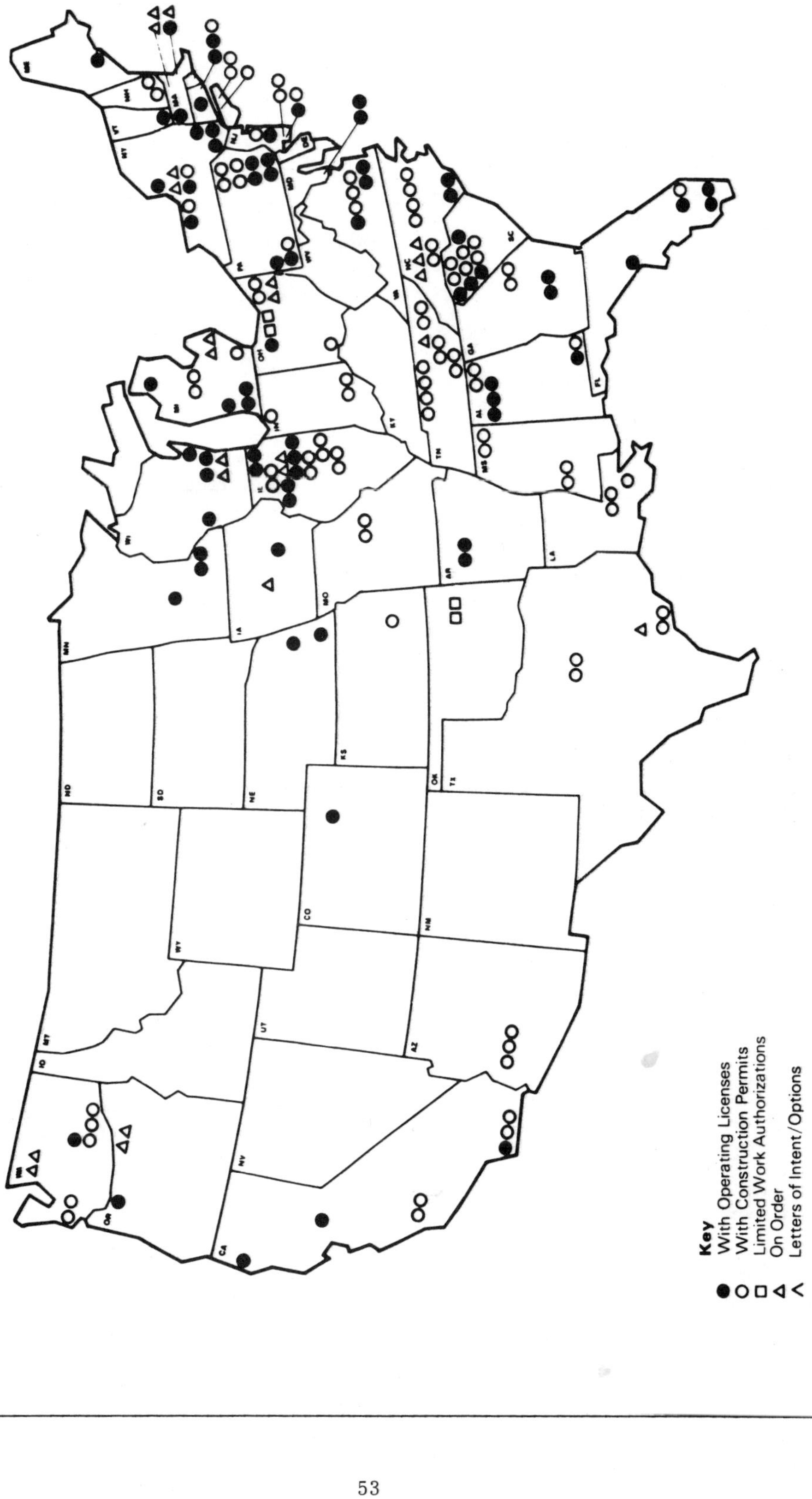

Key
● With Operating Licenses
○ With Construction Permits
□ Limited Work Authorizations
△ On Order
∧ Letters of Intent/Options

Source: Atomic Industrial Forum, Inc.

GEORGIA

Georgia Power Company
?70 Peachtree St.
.tlanta, Georgia 30303 404-522-6060

Operating: Edwin I. Hatch 1 and 2, at Baxley. Boiling water reactors. Net power output: 786 MW and 790 MW, respectively. Commercial operation: December 1975 and September 1979, respectively.

Construction permits: Alvin W. Vogtle 1 and 2, at Waynesboro. Pressurized water reactors. Net power output: 1,100 MW each. Commercial operation: November 1984 and November 1987, respectively.

ILLINOIS

Commonwealth Edison Company
1 First National Plaza
P.O. Box 767
Chicago, Illinois 60690 312-294-4321

Operating: Dresden 2 and 3, at Morris. Boiling water reactors. Net power output: 794 MW each. Commercial operation: July 1970 and November 1971, respectively.

Zion 1 and 2, at Zion. Pressurized water reactors. Net power output: 1,040 MW each. Commercial operation: December 1973 and September 1974, respectively.

Quad Cities 1 and 2, at Cordova. Boiling water reactors. Net power output: 789 MW. Commercial operation: February 1973 and March 1973, respectively.

Construction permits: Braidwood 1 and 2, at Braidwood. Pressurized water reactors. Net power output: 1,120 MW each. Commercial operation: October 1982 and October 1983, respectively.

Byron 1 and 2, at Byron. Pressurized water reactors. Net power output: 1,120 MW each. Commercial operation: October 1982 and October 1983, respectively.

LaSalle 1 and 2, at Seneca. Boiling water reactors. Net power output: 1,078 MW each. Commercial operation: December 1980 and December 1981, respectively.

On order: Carroll County 1 and 2, at Savanna. Pressurized water reactors. Net power output: 1,120 MW each. Commercial operation: October 1990 and October 1991, respectively.

Shut down: Dresden 1, at Morris. Boiling water reactor. Net power output: 207 MW. Shut down October 1978 to upgrade operations. Reoperation: uncertain.



Illinois Power Company
500 S. 27th St.
Decatur, Illinois 62525 217-424-6600

Construction permits: Clinton 1 and 2, at Clinton. Boiling water reactors. Net power output: 950 MW each. Commercial operation: December 1982 and June 1988, respectively.

INDIANA

Northern Indiana Public Service Company
5265 Hohman Ave.
Hammond, Indiana 46320 219-853-5200

Construction permit: Bailly Nuclear 1, at Dunes Acres. Boiling water reactor. Net power output: 644 MW. Commercial operation: December 1987.

Public Service Company of Indiana, Inc.
1000 E. Main St.
Plainfield, Indiana 46168 317-839-9611

Construction permits: Marble Hill 1 and 2, at Madison. Pressurized water reactors. Net power output: 1,130 MW each. Commercial operation: October 1982 and January 1984, respectively.

IOWA

Iowa Electric Light and Power Company
200 First St., S.E.
Cedar Rapids, Iowa 52401 319-398-4411

Operating: Duane Arnold, at Palo. Boiling water reactor. Net power output: 538 MW. Commercial operation: February 1975.

Iowa Power and Light Company
666 Grand Ave.
Des Moines, Iowa 50303 515-281-2900

On order: Vandalia, at Vandalia. Pressurized water reactor. Net power output: 1,270 MW. Commercial operation: deferred indefinitely.

KANSAS

Kansas Gas and Electric Company
201 N. Market St.
Wichita, Kansas 67201 316-261-6611

Construction permit: Wolf Creek, at Burlington. Pressurized water reactor. Net power output: 1,150 MW. Commercial operation: April 1983.

LOUISIANA

Gulf States Utilities Company
285 Liberty Ave.
Beaumont, Texas 77701 713-838-6631

Construction permits: River Bend 1 and 2, at St. Francisville. Boiling water reactors. Net power output: 934 MW each. Commercial operation: September 1984, and indefinite, respectively.

Nuclear Power Plants in the U.S.

LOUISIANA, Cont.:

Louisiana Power & Light Company
142 Delaronde St.
New Orleans, Louisiana 70174 504-366-2345

Construction permit: Waterford 3, at Taft. Pressurized water reactor. Net power output: 1,165 MW. Commercial operation: February 1982.

MAINE

Maine Yankee Atomic Power Company
Edison Dr.
Augusta, Maine 04336 207-623-3521

Operating: Maine Yankee, at Wiscasset. Pressurized water reactor. Net power output: 825 MW. Commercial operation: December 1972.

MARYLAND

Baltimore Gas and Electric Company
P.O. Box 1475
Baltimore, Maryland 21203 301-234-5000

Operating: Calvert Cliffs 1 and 2, at Lusby. Pressurized water reactors. Net power output: 845 MW each. Commercial operation: May 1975 and April 1977, respectively.

MASSACHUSETTS

Boston Edison Company
800 Boylston St.
Boston, Massachusetts 02199 617-424-2400

Operating: Pilgrim 1, at Plymouth. Boiling water reactor. Net power output: 655 MW. Commercial operation: December 1972.

On order: Pilgrim 2, at Plymouth. Pressurized water reactor. Net power output: 1,150 MW. Commercial operation: December 1985.

Northeast Nuclear Energy Company
c/o Northeast Utilities
P.O. Box 270
Hartford, Connecticut 06101 203-623-2421

On order: Montague 1 and 2, at Montague. Boiling water reactors. Net power output: 1,150 MW each. Commercial operation: deferred indefinitely.

Yankee Atomic Electric Company
25 Research Dr.
Westborough, Massachusetts 01581 617-366-9011

Operating: Yankee, at Rowe. Pressurized water reactor. Net power output: 175 MW. Commercial operation: July 1961.

MICHIGAN

Consumers Power Company
212 W. Michigan Ave.
Jackson, Michigan 49201 517-788-0550

Operating: Big Rock Point, at Big Rock Point. Boiling water reactor. Net power output: 63 MW. Commercial operation: March 1963.

Palisades, at South Haven. Pressurized water reactor. Net power output: 740 MW. Commercial operation: December 1971.

Construction permits: Midland 1 and 2, at Midland. Pressurized water reactors. Net power output: 504 MW and 852 MW, respectively. Commercial operation: April 1982 and November 1981, respectively.

Detroit Edison Company
2000 Second Ave.
Detroit, Michigan 48226 313-237-8000

Construction permit: Enrico Fermi 2, at Lagoona Beach. Boiling water reactor. Net power output: 1,093 MW. Commercial operation: March 1982.

On order: Greenwood 2 and 3, in St. Clair County. Pressurized water reactors. Net power output: 1,264 MW each. Commercial operation: September 1990 and September 1992, respectively.

Indiana & Michigan Electric Company
2101 Spy Run Ave.
Fort Wayne, Indiana 46801 219-422-3456

Operating: Donald C. Cook 1 and 2, at Bridgman. Pressurized water reactors. Net power output: 1,054 MW and 1,100 MW, respectively. Commercial operation: August 1975 and July 1978, respectively.

MINNESOTA

Northern States Power Company (Minnesota)
414 Nicollet Mall
Minneapolis, Minnesota 55401 612-330-5500

Operating: Monticello, at Monticello. Boiling water reactor. Net power output: 545 MW. Commercial operation: June 1971.

Prairie Island 1 and 2, at Red Wing. Pressurized water reactors. Net power output: 530 MW each. Commercial operation: December 1973 and December 1974, respectively.

MISSISSIPPI

Mississippi Power & Light Company
Electric Building
Jackson, Mississippi 39205 601-969-2311

Construction permits: Grand Gulf 1 and 2, at Port Gibson. Boiling water reactors. Net power output: 1,250 MW each. Commercial operation: April 1981 and January 1984, respectively.

MISSISSIPPI, Cont.:

Tennessee Valley Authority (TVA)
Commerce Building
400 Commerce Ave.
Knoxville, Tennessee 37902 615-632-2101

Construction permits: Yellow Creek 1 and 2, in Tishimingo County. Pressurized water reactors. Net power output: 1,205 MW. Commercial operation: May 1985 and April 1988, respectively.

MISSOURI

Union Electric Company
1901 Gratiot St.
St. Louis, Missouri 63103 314-621-3222

Construction permits: Callaway 1 and 2, in Callaway County. Pressurized water reactors. Net power output: 1,150 MW each. Commercial operation: October 1982 and April 1987, respectively.

NEBRASKA

Nebraska Public Power District
P.O. Box 499
Columbus, Nebraska 68601 402-564-8561

Operating: Cooper, at Brownville. Boiling water reactor. Net power output: 778 MW. Commercial operation: July 1974.

Omaha Public Power District
1623 Harnay St.
Omaha, Nebraska 68102 402-536-4000

Operating: Fort Calhoun 1, at Fort Calhoun. Pressurized water reactor. Net power output: 457 MW. Commercial operation: June 1974.

NEW HAMPSHIRE

Public Service Company of New Hampshire
1000 Elm St.
P.O. Box 330
Manchester, New Hampshire 03105 603 669-4000

Construction permits: Seabrook 1 and 2, at Seabrook. Pressurized water reactors. Net power output: 1,194 MW each. Commercial operation: April 1983 and February 1985, respectively.

NEW JERSEY

Jersey Central Power & Light Company
Madison Ave. at Punch Bowl Rd.
Morristown, New Jersey 07960 201-455-8200

Operating: Oyster Creek, in Lacey Township. Boiling water reactor. Net power output: 650 MW. Commercial operation: December 1969.

Construction permit: Forked River 1, in Lacey Township. Pressurized water reactor. Net power output: 1,168 MW. Commercial operation: indefinite.

Nuclear Power Plants in the U.S.

Public Service Electric and Gas Company
80 Park Place
Newark, New Jersey 07101 201-430-7000

Operating: Salem 1, at Salem. Pressurized water reactor. Net power output: 1,090 MW. Commercial operation: June 1977.

Construction permits: Salem 2, at Salem. Pressurized water reactor. Net power output: 1,115 MW. Commercial operation: May 1980.

Hope Creek 1 and 2, in Salem County. Boiling water reactors. Net power output: 1,067 MW each. Commercial operation: May 1984 and May 1986, respectively.

NEW YORK

Consolidated Edison Company of New York, Inc.
4 Irving Place
New York, New York 10003 212-460-4600

Operating: Indian Point 2, at Buchanan. Pressurized water reactor. Net power output: 873 MW. Commercial operation: August 1973.

Shut down: Indian Point 1, at Buchanan. Pressurized water reactor. Net power output: 265 MW. Reoperation: no decision yet made.

Long Island Lighting Company (LILCO)
250 Old Country Rd.
Mineola, New York 11501 516-228-2890

Construction permits: Jamesport 1 and 2, at Riverhead. Pressurized water reactors. Net power output: 1,150 MW. Commercial operation: July 1988 and July 1990, respectively.

Shoreham, at Brookhaven. Boiling water reactor. Net power output: 854 MW. Commercial operation: December 1981.

New York State Electric & Gas Corporation
4500 Vestal Parkway East
Binghamton, New York 13902 607-729-2551

On order: New Haven 1 and 2, at New Haven. Pressurized water reactors. Net power output: 1,250 MW. Commercial operation: May 1993 and May 1994, respectively.

Niagara Mohawk Power Corporation
300 Erie Blvd., West
Syracuse, New York 13202 315-474-1511

Operating: Nine Mile Point 1, at Oswego. Boiling water reactor. Net power output: 620 MW. Commercial operation: December 1969.

Construction permit: Nine Mile Point 2, at Oswego. Boiling water reactor. Net power output: 1,080 MW. Commercial operation: October 1984.

Nuclear Power Plants in the U.S.

NEW YORK, Cont.:

Power Authority of the State of New York
10 Columbus Circle
New York, New York 10019 212-397-6200

Operating: Indian Point 3, at Buchanan. Pressurized water reactor. Net power output: 965 MW. Commercial operation: August 1976.

James A. FitzPatrick, at Scriba. Boiling water reactor. Net power output: 821 MW. Commercial power operation: July 1975.

Rochester Gas and Electric Corporation
89 East Ave.
Rochester, New York 14649 716-546-2700

Operating: Robert E. Ginna, at Rochester. Pressurized water reactor. Net power output: 470 MW. Commercial operation: July 1970.

Construction permit: Sterling, at Sterling. Pressurized water reactor. Net power output: 1,150 MW. Commercial operation: May 1988.

NORTH CAROLINA

Carolina Power & Light Company
Center Plaza Building
411 Fayetteville St.
Raleigh, North Carolina 27602 919-836-6111

Operating: Brunswick 1 and 2, at Southport. Boiling water reactors. Net power output: 821 MW each. Commercial operation: March 1977 and November 1975, respectively.

Construction permits: Shearon Harris 1, 2, 3 and 4, at New Hill. Pressurized water reactors. Net power output: 900 MW each. Commercial operation: March 1984, March 1986, March 1981 and March 1989, respectively.

Duke Power Company
422 S. Church St.
Charlotte, North Carolina 28242 704-373-4011

Construction permits: William McGuire 1 and 2, at Cowans Ford Dam. Pressurized water reactors. Net power output: 1,180 MW each. Commercial operation: August 1980? and April 1982, respectively.

On order: Thomas L. Perkins 1, 2, and 3, in Davie County. Pressurized water reactors. Net power output: 1,280 MW each. Commercial operation: deferred indefinitely.

OHIO

Central Area Power Coordination Group (CAPCO)
Cleveland Electric Illuminating Company (operating utility)
55 Public Square
Cleveland, Ohio 44101 216-622-9800

Construction permits: Perry 1 and 2, at North Perry. Boiling water reactors. Net power output: 1,205 MW each. Commercial operation: May 1983 and May 1985, respectively.

Central Area Power Coordination Group (CAPCO)
Ohio Edison Company (operating utility)
76 S. Main St.
Akron, Ohio 44308 216-384-5100

On order: Erie 1 and 2, at Berlin Heights. Pressurized water reactors. Net power output: 1,267 MW each. Commercial operation: April 1986 and April 1988, respectively (may be delayed up to 3 years).

Central Area Power Coordination Group (CAPCO)
Toledo Edison Company (operating utility)
300 Madison Ave.
Toledo, Ohio 43652 419-259-5000

Operating: Davis-Besse 1, at Oak Harbor. Pressurized water reactor. Net power output: 906 MW. Commercial operation: November 1977.

Limited work authorization: Davis-Besse 2 and 3, at Oak Harbor. Pressurized water reactors. Net power output: 906 MW each. Commercial operation: April 1985 and April 1987, respectively (may be delayed up to 3 years).

Cincinnati Gas & Electric Company
139 E. Fourth St.
Cincinnati, Ohio 45202 513-381-2000

Construction permit: William H. Zimmer, at Moscow. Boiling water reactor. Net power output: 810 MW. Commercial operation: January 1980 (initial projection).

OKLAHOMA

Public Service Company of Oklahoma
212 E. Sixth St.
Tulsa, Oklahoma 74119 918-583-3611

Limited work authorizations: Black Fox 1 and 2, at Inola. Boiling water reactors. Net power output: 1,150 MW each. Commercial operation: July 1985 and July 1988, respectively.

OREGON

Portland General Electric Company
121 S.W. Salmon St.
Portland, Oregon 97204 503-226-8333

Operating: Trojan, at Rainier. Pressurized water reactor. Net power output: 1,130 MW. Commercial operation: May 1976.

On order: Pebble Springs 1 and 2, at Arlington. Pres-

OREGON, Cont.:

surized water reactors. Net power output: 1,260 MW each. Commercial operation: October 1988 and October 1990, respectively.

PENNSYLVANIA

Central Area Power Coordination Group (CAPCO)
Duquesne Light Company (operating utility)
435 Sixth Ave.
Pittsburgh, Pennsylvania 15219 412-456-6000

Operating: Beaver Valley 1, at Shippingport. Pressurized water reactor. Net power output: 852 MW. Commercial operation: October 1976.

Construction permit: Beaver Valley 2, at Shippingport. Pressurized water reactor. Net power output: 852 MW. Commercial operation: May 1984.

Metropolitan Edison Company
2800 Pottsville Pike
Muhlenberg Township
Berkshire County, Pennsylvania 215-929-3601

Shut down: Three Mile Island 1 and 2, in Londonderry Township. Pressurized water reactors. Net power output: 819 MW and 906 MW, respectively. Shut down subsequent to March 1979 accident. Reoperation: pending NRC consideration of accident.

U.S. Department of Energy
Duquesne Light Company (operating utility)
435 Sixth Ave.
Pittsburg, Pennsylvania 15219 412-456-6000

Operating: Shippingport, at Shippingport. Light water breeder reactor. Net power output: 60 MW. Commercial operation: December 1957.

Pennsylvania Power & Light Company
2 N. Ninth St.
Allentown, Pennsylvania 18101 215-821-5151

Construction permits: Susquehanna 1 and 2, at Berwick. Boiling water reactors. Net power output: 1,050 MW each. Commercial operation: January 1982 and January 1983, respectively.

Philadelphia Electric Company
2301 Market St.
Philadelphia, Pennsylvania 19101 215-841-4000

Operating: Peach Bottom 2 and 3, in Peach Bottom Township. Boiling water reactors. Net power output: 1,065 MW each. Commercial operation: July 1974 and December 1974, respectively.

Construction permits: Limerick 1 and 2, in Limerick Township. Boiling water reactors. Net power output: 1,055 MW each. Commercial operation: April 1985 and April 1987, respectively.



SOUTH CAROLINA

Carolina Power & Light Company
Center Plaza Building
411 Fayetteville St.
Raleigh, North Carolina 27602 919-836-6111

Operating: H.B. Robinson 2, at Hartsville. Pressurized water reactor. Net power output: 700 MW. Commercial operation: March 1971.

Duke Power Company
422 S. Church St.
Charlotte, North Carolina 28242 704-373-4011

Operating: Oconee 1, 2 and 3, at Lake Keowee. Pressurized water reactors. Net power output: 887 MW each. Commercial operation: July 1973, September 1974, and December 1974, respectively.

Construction permits: Catawba 1 and 2, in York County. Pressurized water reactors. Net power output: 1,145 MW each. Commercial operation: July 1983 and January 1985, respectively.
Cherokee 1, 2 and 3, in Cherokee County. Pressurized water reactors. Net power output: 1,280 MW each. Commercial operation: January 1985, January 1987, and January 1989, respectively.

South Carolina Electric & Gas Company
328 Main St.
P.O. Box 764
Columbia, South Carolina 29218 803-799-1234

Construction permit: Virgil C. Summer 1, at Parr. Pressurized water reactor. Net power output: 900 MW. Commercial operation: December 1980 (projected date).

TENNESSEE

Tennessee Valley Authority (TVA)
Commerce Building
400 Commerce Ave.
Knoxville, Tennessee 37902 615-632-2101

Construction permits: Hartsville A-1 and A-2, at Hartsville. Boiling water reactors. Net power output: 1,205 MW each. Commercial operation: July 1986 and July 1987, respectively.
Hartsville B-1 and B-2, at Hartsville. Boiling water reactors. Net power output: 1,205 MW each. Commercial operation: June 1989 and June 1990, respectively.
Phipps Bend 1 and 2, at Surgoinsville. Boiling water reactors. Net power output: 1,220 MW each. Commercial operation: August 1984 and August 1989, respectively.
Sequoyah 1 and 2, at Daisy. Pressurized water reactors. Net power output: 1,140 MW each. Commercial operation: June 1980 and June 1981, respectively.
Watts Bar 1 and 2, at Spring City, Pressurized water reactors. Net power output: 1,165 MW each. Commercial operation: September 1981 and June 1982, respectively.

Nuclear Power Plants in the U.S.

TENNESSEE, Cont.:

On order: Clinch River Breeder Reactor Plant, at Oak Ridge. Liquid metal fast breeder reactor. Net power output: 350 MW. Commercial operation: September 1988 (subject to resolution of national policy debate).

TEXAS

Houston Lighting & Power Company
P.O. Box 1700
Houston, Texas 77001 713-228-9211

On order: Allens Creek, at Wallis. Boiling water reactor. Net power output: 1,150 MW. Commercial operation: February 1987.

South Texas Project
Houston Lighting & Power Company (Project manager)
P.O. Box 1700
Houston, Texas 77001 713-228-9211

Construction permits: South Texas Project 1 and 2, in Matagorda County. Pressurized water reactors. Net power output: 1,250 MW each. Commercial operation: February 1984 and February 1986, respectively.

Texas Utilities Generating Company
2001 Bryan Tower
Dallas, Texas 75201 214-653-4600

Construction permits: Comanche Peak 1 and 2, in Somervell County. Pressurized water reactors. Net power output: 1,150 MW each. Commercial operation: January 1981 and January 1983, respectively.

VERMONT

Vermont Yankee Nuclear Power Corporation
77 Grove St.
Rutland, Vermont 05701 802-773-2711

Operating: Vermont Yankee, at Vernon. Boiling water reactor. Net power output: 514 MW. Commercial operation: November 1972.

VIRGINIA

Virginia Electric and Power Company
1 James River Plaza
P.O. Box 26666
Richmond, Virginia 23261 804-771-3000

Operating: Surry 1 and 2, at Gravel Neck. Pressurized water reactors. Net power output: 822 MW each. Commercial operation: December 1972 and May 1973, respectively.
North Anna 1, at Mineral. Pressurized water reactor. Net power output: 907 MW. Commercial operation: June 1978.

Construction permits: North Anna 2, 3 and 4, at Mineral. Pressurized water reactors. Net power output: 907 MW each. Commercial operation: July 1980 (projected date), November 1986, and December 1987, respectively.

WASHINGTON

Puget Sound Power and Light Company
Puget Power Building
Bellevue, Washington 98009 206-454-6363

On order: Skagit 1 and 2, at Sedro Woolley. Boiling water reactors. Net power output: 1,288 MW each. Commercial operation: September 1988 and September 1990, respectively.

U.S. Department of Energy
Washington Public Power Supply System (distributor)
3000 George Washington Way
P.O. Box 968
Richland, Washington 99352 509-375-5000

Operating: Hanford - N, at Richland. Graphite-moderated reactor. Net power output: 800 MW. Commercial operation: September 1966.

Washington Public Power Supply System
3000 George Washington Way
P.O. Box 968
Richland, Washington 99352 509-375-5000

Construction permits: WPPSS 1, 2, 3, 4 and 5, at Richland and Satsop. Pressurized water reactors, except for 2, which is a boiling water reactor. Net power output: 1,267 MW, 1,093 MW, 1,240 MW, 1,267 MW, and 1,240 MW, respectively. Commercial operation: December 1983, September 1981, December 1984, June 1985, and June 1986, respectively.

WISCONSIN

Dairyland Power Cooperative
2615 East Ave., South
LaCrosse, Wisconsin 54601 608-788-4000

Operating: LaCrosse, at Genoa. Boiling water reactor. Net power output: 50 MW. Commercial operation: November 1969.

Wisconsin Electric Power Company
231 W. Michigan St.
P.O. Box 2046
Milwaukee, Wisconsin 53201 414-277-2345

Operating: Point Beach 1 and 2, at Two Creeks. Pressurized water reactors. Net power output: 497 MW each. Commercial operation: December 1970 and October 1972, respectively.

On order: Haven 1 and 2, at Haven. Pressurized water reactors. Net power output: 900 MW each. Commercial operation: June 1989, and indefinite (licensing review suspended), respectively.

WISCONSIN, Cont.:

Wisconsin Public Service Corporation
700 N. Adams St.
Green Bay, Wisconsin 54301 414-433-1598

Operating: Kewaunee, in Carlton Township. Pressurized water reactor. Net power output: 535 MW. Commercial operation: June 1974.

# Chapter 5.
# International Organizations

INTERNATIONAL INTER-GOVERNMENTAL ORGANIZATIONS:

Euratom/Communauté Européene de l'Energie Atomique (CEEA)/Europaeische Atomgemeinschaft (EAG) Comunita Europea dell'Energia Atomica (CEEA)/ Europese Gemeenschap voor Atomenergie (EGA)
200 rue de la Loi
1040 Brussels
Belgium 735-8040

Founded 1958. Comprised of nine member states of the European Community. Concerned with the development of nuclear research and nuclear health protection standards. Cooperates in OECD Nuclear Energy Agency research projects. Pub: Journal Officiel; Euro Spectra, q.

European Atomic Energy Society (EAES)/Société Européenne d'Energie Atomique (SEEA)
c/o Risø National Laboratory
DK-4000 Roskilde
Denmark 03-35 51 9

Created in 1954 to encourage cooperation in atomic energy research and its peaceful applications. Membership is composed of national commissions from 13 different countries.

The European Community
Commission of the European Communities
rue de la Loi 200
1040 Brussels
Belgium

Includes the European Economic Community, the European Coal and Steel Community, Euratom, and related agencies. The treaty establishing Euratom contained enabling provisions for setting Community standards on protection against ionizing radiation. Though not responsible for broad environmental protection measures, the Community has dealt with problems associated with contamination by radioactive wastes.

Members of the Community are: Belgium, Denmark, Federal Republic of Germany, France, Ireland, Italy, Luxembourg, Netherlands, and the United Kingdom. Information Services are maintained in member nations, and in Montevideo, New York, and Washington, D.C. Pub: Bulletin of the European Communities, m.; Community Energy policy, book; Progress Report on the Radiation Protection Programme; Nuclear science and technology: Catalogue and classification of technical safety standards, rules and regulations for nuclear power reactors and nuclear fuel cycle facilities, book; Community nuclear safety code, study. (Publications available from the Press Offices in member countries.)

European Company for the Chemical Processing of Irradiated Fuels/Société Européenne pour le Traitement Chimique des Combustibles Irradies (EUROCHEMIC)
2400 Mol-Donk
Belgium 014-31 28 61

Established in 1957. Thirteen member countries. Supports research and industrial activity related to the processing of irradiated fuels.

European Organization for Nuclear Research/Conseil Européen de la Recherche Nucléaire (CERN)
Meyrin
1211 Geneva 23
Switzerland 41 98 11

Established 1954. Members are governments of 12 countries. Promotes collaboration in pure and fundamental nuclear research. Pub: CERN Courrier, m.

Food and Agriculture Organization of the United Nations (FAO)
Viale delle Terme di Caracalla
00100 Rome
Italy

Established 1945. General concerns for world agriculture and rural populations. With the IAEA, FAO convenes a joint Panel of Experts on Radiation Contamination of Food. Pub: FAO Review, m.

Inter-American Nuclear Energy Commission (IANEC)/ Comision Interamericana de Energia Nuclear (CIEN)
c/o Organization of American States (OAS)
Pan American Union
Washington, D.C. 20006
U.S.A.

The OAS was established in 1890 to strengthen peace and security in the Americas and promote, through cooperation, general economic, social and cultural development. It has 24 members from both North and South America.

Members of IANEC represent member states of the OAS. IANEC was established in 1959 to assist members in the development of coordinated plans for research and training, and exchange of technical and scientific data on peaceful uses of atomic energy, including public health aspects.

International Atomic Energy Agency (IAEA)
11 Kaerntnerring
P.O. Box 590
1011 Vienna
Austria

The IAEA, a Specialized (autonomous) Agency of the United Nations, was created in 1957 to "seek to accelerate and enlarge the contribution of atomic energy to peace, health and prosperity throughout the world" and to "ensure so far as it is able, that assistance provided by it or at its request or under its supervision or control is not used in such a way as to further any military purpose."

The IAEA has more than 100 member states, to which it gives advice and provides technical assistance in the following areas relating to nuclear power: water desalination applications, health and safety, radioactive waste management, legal aspects of nuclear power use, and prospecting and exploiting nuclear raw materials. The IAEA promotes the use of radiation and isotopes in agriculture, industry, medicine, biology and hydrology.

The Agency is particularly concerned with all aspects of radiation safety. It has established regulations for safe transport of radioactive materials by rail, road, sea, and air, which have been adopted as legal standards by many governments. They have also been included in conventions and recommendations of nearly all international organizations concerned with transport.

IAEA also advises developing countries on safe nuclear waste treatment and disposal techniques. It has adopted codes for the safe operation of nuclear power plants and research reactors, and acted as counsel to several countries with regards to nuclear reactor siting. The Agency has developed a set of recommendations on the use of ports by nuclear-powered merchant ships; it has determined principles for limiting the introduction of nuclear wastes into the sea; and, it maintains an international register of releases of radioactive material to any sector of the environment beyond national jurisdictions.

A major function of the IAEA is to establish and administer safeguards to ensure that nuclear materials and equipment intended for peaceful uses are not diverted to military purposes. Under the Treaty on the Non-Proliferation of Nuclear Weapons, which took effect in 1970, IAEA is responsible for applying safeguards with respect to all nuclear materials in all the peaceful nuclear activities of all non-nuclear-weapon states which ratified it. In addition, any country which receives assistance from the Agency must accept its safeguards for that particular assistance project. Safeguards are also applied on request.

Pub: IAEA Bulletin, bi-m; Nuclear Fusion, bi-m.; Atomic Energy Review, q.; proceedings, technical reports, and bibliographies, complete catalogue available.

The International Nuclear Fuel Cycle Evaluation (INFCE)

Initiated by the U.S. government in 1977 "to define the technical areas in which nations might cooperate to prevent the use of commercial nuclear systems for warlike purposes"; in other words, to determine international controls to be placed on radioactive fuels and nuclear technologies (especially plutonium and breeder reactors) that might be used to make bombs. The INFCE involved 66 countries and 4 international organizations which met in Vienna, worked in eight specialized groups for about two years, and produced a final report issued on February 26, 1980.

The report did not fulfill the U.S.'s expectations. It contained few hard, specific points of agreement among the participating countries. It contained the following: a lower figure for international demand for nuclear power than previously used, and the conclusion that uranium supplies will be adequate through the year 2000 (implying that fuel recycling or breeder reactors are not needed urgently); a finding that existing and planned fuel recycling and enrichment facilities will be adequate to supply the nuclear industry until 1990 at least, and that further investments in such facilities would be uneconomical; a statement that there are safe ways of coping with the problem of nuclear waste storage (specific ways await conclusion of a study overseen by the IAEA); a proposal by the INFCE itself to undertake a study to develop an international program for auditing and controlling civilian plutonium inventories; and a statement of support for work in the U.S. and Europe to lower the grade of uranium used in research reactors, in order to reduce the threat of diversion to military ends (INFCE had no new ideas on how to fund the fuel conversion effort).

Nordisk Institut for Teoretisk Atomfysik (Nordic Institute for Theoretical Nuclear Physics) (NORDITA)
Niels Bohr Institute
University of Copenhagen
Copenhagen
Denmark

Joint Scandinavian research group.

Organismo Para La Proscripcion de Las Armas Nucleares en La America Latina (Organization for the Prohibition of Nuclear Weapons in Latin America) (OPANAL)
Témistocles 78, Col. Polanco
México 5
D.F., México 250-62-22

Responsible for holding consultations among member states on matters relating to the purposes, measures and procedures set forth in the Treaty of Tlatelolco (prohibiting nuclear weapons), and for supervising compliance with the obligations arising from such consultations. Principal organs of OPANAL include the General Conference, the Council, and the Secretariat.

Joint Institute for Nuclear Research

See under U.S.S.R., p. 98.

OECD Nuclear Energy Agency (NEA)/Agence de l'OCDE pour l'Energie Nucléaire (AEN)
38 bd Suchet
75016 Paris
France

The OECD (Organisation for Economic Co-operation and Development) has 24 member states. It is concerned with sustainable growth and rising standards of living in member states, with economic growth in developing countries, and with the expansion of world trade "on a multi-lateral non-discriminatory basis in accordance with international obligations."

The NEA is an agency of the OECD, established in 1972 to replace the European Nuclear Energy Agency (ENEA). It is concerned to develop production and uses of nuclear energy for peaceful purposes through joint undertakings and common services, scientific and technical cooperation, studies of energy requirements, and establishment of uniform legislation governing nuclear safety. NEA maintains a committee on Radiological Protection and Public Health. Its members include governments of 19 countries.

Scandinavian Committee for Power Supply (NORDEL)
Imatran Voima Osakeyhtiö
P.O. Box 138
SF-00101 Helsinki 10
Finland

Nordic cooperative organization, established on the recommendation of the Nordic Council of 1959. Promotes cooperation among member states in the production, distribution and consumption of electrical energy. Collaboration on both daily operations and long-term planning of power systems and nuclear operations. The Secretariat rotates every three months among the Nordic countries.

The United Nations
Public Inquiries Unit
New York, New York 10017 212-754-7721

Within the United Nations system, various organs perform work related to the nuclear power issue. Three Specialized Agencies have specific functions relating to nuclear power: the International Atomic Energy Agency, the Food and Agriculture Organization, and the World Health Organization. The World Meteorological Organization also performs some research related to nuclear matters. For descriptions of these Agencies, see elsewhere in this chapter.

World Health Organization (WHO)
1211 Geneva 27
Switzerland

Created in 1946 with the goal of attaining for all peoples "the highest possible level of health," that is, "a state of complete physical, mental and social well-being and not merely the absence of disease or infirmity." WHO is actively concerned with radiation from the standpoint of human health. Its Secretariat is organized along functional lines, one of which is Radiation Health. WHO also has an expert advisory panel on Radiation. Pub: World Health, m.; Health Implications of Nuclear Power Production, book; Protection of the Public in the Event of Radiation Accidents; Mental Health Aspects of the Peaceful Uses of Atomic Energy, report (1958); other publications, catalogue available.

World Meteorological Organization (WMO)
41, avenue Giuseppe-Motta
1211 Geneva 20
Switzerland

Created in 1947 as a Specialized Agency of the United Nations. Its mission includes "the application of meteorology to aviation, shipping, and other maritime affairs, agriculture, water problems, and other human activities." One of WMO's activities is to maintain a global network to measure isotopes in precipitation, in collaboration with the IAEA. Sixty-seven countries, with more than 100 stations, participate in this network. Pub: WMO Bulletin, q. (general).

## INTERNATIONAL NONGOVERNMENTAL ORGANIZATIONS:

European Nuclear Society/Europaeische Kernenergie-Gesellschaft/Société Européenne de l'Energie Nucléaire (ENS)
Legal Seat
6, rue d'Italie
P.O. Box 737
CH 1211 Geneva 3
Switzerland

Secretariat:
Baerenplatz 2
P.O. Box 2613
CH 3001 Bern
Switzerland

Scholarly society whose membership includes 15 national societies and, altogether, more than 9,000 scientists and engineers. ENS aims "to promote and to contribute to the advancement of science and engineering in the field of the peaceful uses of nuclear energy by all suitable means and in particular by: a) fostering and co-ordinating the activities of the member organizations. b) encouraging exchanges between the member organizations. c) encouraging the exchange of scientists and engineers between countries. c) disseminating information." ENS, in cooperation with its national member societies or the American Nuclear Society, sponsors several topical meetings in Europe and the U.S. each year. Pub: ENS Newsletter, bi-m.; Nuclear Technology, published jointly with the American Nuclear Society.

Friends of the Earth International (FOE)
124 Spear St.
San Francisco, California 94105
U.S.A. 415-495-4770

World-wide affiliation of national Friends of the Earth. Seeks interchange and cooperation on an international level, and exchange of information. FOE's energy policy calls for a gradual halt to energy growth rates and a moratorium on nuclear fission reactors, with development of alternative, "soft" energy paths. The Friends of the Earth Foundation (located at the same address) engages in scientific and educational work. FOE has principal U.S. offices in New York, Seattle, Washington, and Washington, D.C. It has sister organizations in Australia, Austria, Canada, El Salvador, France, the Federal Republic of Germany, Greece, Ireland, Italy, Japan, Malaysia, Mexico, the Netherlands, New Zealand, Portugal, Spain, Sweden, Switzerland, Thailand, the United Kingdom, and Yugoslavia.

Pub: Not Man Apart, m.

Institute for World Order
The World Order Models Project (WOMP)
777 United Nations Plaza
New York, New York 10017 212-490-0010

The Institute was founded in 1961 to promote practical alternatives to war, social and economic justice, and ecological breakdown. WOMP is an individual membership association engaged in research, education, dialogue, and action to promote a just world order. It held its Sixteenth Annual Conference July 1980 in Lisbon, and at that time drafted a policy document entitled, "Denuclearization For A Just World: The Failure of Non-Proliferation." WOMP is dedicated to finding practical and equitable solutions to the problems posed by nuclear power and nuclear weapons. Pub: Working papers, include "Nuclear Power and World Order: Why Denuclearization," an argument against the use of nuclear technology for any purpose whatsoever.

International Commission on Radiological Protection (ICRP)
Clifton Ave.
Sutton, Surrey
United Kingdom SM2 5PU 01-642 4680

Established in 1928 by the Second International Congress of Radiology as the International X-Ray and Radium Protection Committee; name and organizational structure changed in 1950 to provide "general guidance on the widespread use of radiation sources caused by rapid developments in the field of nuclear energy." Maintains contact with medical radiology and the medical profession in general. ICRP policy is "to consider the fundamental principles upon which appropriate radiation protection measures can be based while leaving to the various national protection bodies the responsibility of formulating the specific advice, codes of practice, or regulations that are best suited to the needs of their individual countries."

The Commission's work for the last four-year period (1977-1981) has been performed by four committees: Radiation Effects, Secondary Limits, Protection in Medicine, and Application of the Commission's Recommendations. ICRP maintains official relationships with the World Health Organization and the International Atomic Energy Agency, and close working relationships with the U.N. Scientific Committee on the Effects of Atomic Radiation, the International Labour Office, the U.N. Environment Programme, the Nuclear Energy Agency, and the European Economic Community. Pub: Radionuclide Release into the Environment: Assessment of Doses to Man, report (1979); Limits for Intake of Radionuclide by Workers, report (1979); Annals of the ICRP, irr.; complete list of publications available.

International Conference for Coordination of the Anti-Nuclear Movement/Internationale Koordinationskonferenz der Anti-Atomenergie-Bewegung/Conference Internationale de Coordination du Mouvement Anti-Nucléaire
P.O. Box 231
CH-4015 Basel
Switzerland 061-38 63 26

First called in March 1978, by the National Coordination of the Swiss Anti-Nuclear Organizations (which see). Works for a shift in the balance of power between corporate and government interests on the one hand, and anti-nuclear interests on the other. Effort involves: coordination of a strong national movement; international coordination (such as demonstrations held June 1979 at Whitsuntide); an appeal to labor unions and other members of the working class to join the movement. Calls for a moratorium on all nuclear activities, as well as popular referenda on the question of nuclear development. Organizes a series of International Conferences as a forum for exchange of ideas. The Fifth Conference was held September 1980 in Basel. Pub: Information bulletins; minutes of the International Conferences, annual.

The International Institute for Strategic Studies
23 Tavistock St.
London WC2E 7NQ
United Kingdom

Founded 1958 by British analysts, academics, politicians, journalists, churchmen to study security in the nuclear age. In 1964, the Institute went international. It has more than 60 member countries, and operates as a research and information center, and a catalyst to discussion and debate in the belief that "rationality makes a major contribution to security." Pub: The Military Balance, annual.

International Resistance to the War and International Movement of Reconciliation/International de Resistants à la Guerre et Mouvement Internationale de la Reconciliation (IRG/MIR)
Namur
rue Haute Marcelle, 11
B-5000 Belgium 081-22.46.16

Opposes nuclear power as it is linked with nuclear weapons proliferation. Main concern is with disarmament. Pub: Nonviolence et Société, m.; Une Autre Defense, m.

International Symposium on the Scientific Basis for Nuclear Waste Management
Materials Research Society
102C Materials Research Laboratory
University Park, Pennsylvania 16802

Annual meeting, with presentation of papers to provide a technical exchange of information between various disciplines (geology, management, storage) involved in nuclear waste management. Pub: Proceedings, annual.

Nuclear Free Pacific Conference/1980
c/o Pacific Concerns Resource Center
1212 University Ave.
Honolulu, Hawaii 16826 808-947-8403

Third conference of its kind (first two held in 1975 and 1978) for nations in the Pacific. Representatives at the conference came from the Cook Islands, Fiji, Tahiti, New Caledonia, Vanuatu (New Hebrides), Tonga, Guam, the Marshall Islands, Palau, Tinian and Truk; also Japan, the Philippines, the Netherlands; and indigenous peoples including Australian Aborigines, New Zealand Maories, and American Indians from the U.S. and Canada were represented also. The Conference reaffirmed its position, taken at previous meetings, of support for native peoples working to "regain control over lands alienated from them through the workings of imperialism and militarism; to halt cultural genocide; to stop the uranium mining on lands of indigenous peoples; to end the exploitation of cultures and peoples; to win autonomy or independence from foreign control; to build greater unity among the aboriginal or indigenous peoples of the Pacific toward ending the oppression of indigenous peoples and gaining a Pacific that is not only nuclear-free, but is also oppression free."

The Conference established the following new mechanisms to implement its stand: a "RIMPAC" network to address joint military maneuvers by the U.S., Australia, New Zealand, Canada, and Japan in the Pacific Rim; a Pacific Trade Union Forum to deal with dangers of the nuclear industry in the Pacific; a Trident network to coordinate protest activities against the nuclear submarines; and a task force to monitor exports of U.S. nuclear waste and power plants to the Pacific. The Conference also established the Pacific Concerns Resource Center (see below). Pub: Follow Up Report, available from the Pacific Concerns Resources Center.

Pacific Concerns Resource Center
1212 University Ave.
Honolulu, Hawaii 96826 808-947-8403

Created as a locus for Pacific solidarity by the Nuclear Free Pacific Conference (see above) held in Hawaii May 10-18, 1980. Mandated as a clearinghouse for information on Pacific nuclear, land rights, and independence issues. Perform's research, produces slide shows, publications and other resources. Pub: Newsletter, 8 times/yr.; Nuclear Free Pacific Conference/1980 Follow Up Report.

Stockholm International Peace Research Institute (SIPRI)
Sveavaegen 166
S-113 46 Stockholm
Sweden 08-15 09 40

Independent research institute concerned with peace and conflict, disarmament and arms regulation. Financed by the Swedish Parliament, but otherwise international in constitution. Pub: SIPRI Yearbook, annual analysis of the world arms race, including information on nuclear weapons and attempts at reconciliation.

Union for the Coordination of the Production and Transport of Electric Power (UCPTE)
Utrechtseweg 310
6812 AR Arnhem
The Netherlands 85 457057

Founded 1951. Eight member countries in Europe. Concerned with the energy situation, especially electric power, in member countries and other countries nearby, and with the extension of existing energy lines and power stations. UCPTE is divided into three working groups: Operating Problems (which studies such things as the use of computers in international energy programs), Thermal Power Plants (studies thermal power and reviews various countries' nuclear power policies and the general status of nuclear power), and Hydroelectric Production. UCPTE's Secretariat prepares forecasts of export and import possibilities for electrical energy in the member countries, and information on the international exchange of power peak and off-peak hours. Pub: Annual Report; UCPTE-Quarterly Reports (general energy situation coverage).

War Resisters International (WRI)
55 Dawes St.
London SE 17 1EL
United Kingdom 01-703 71 89

International association of war resisters and pacifists. Opposes nuclear energy programs; works for disarmament. Pub: WRI Newsletter, 6 times/yr.

World Energy Conference (WEO)
Central Office, International Executive Council
34 St. James's St.
London SW1A 1HD
United Kingdom 01-930 3966

Aims "to promote the development and the peaceful use of energy resources to the greatest benefit of all, both nationally and internationally." Considers the production, transportation, transformation, utilization of energy resources; puts energy consumption in the broad context of economic activity; assists and attends meetings of other energy organizations and acts as an information center and clearinghouse; works to promote its broad aims as stated above. Collects and publishes data on energy; holds conferences for those concerned with energy (11 conferences in all have been held).

WEO is composed of an International Executive Council, National Committees, Programme and Administrative Committees, a Conservation Commission, and various other committees. The conferences feature a number of "Round Tables," including in 1980 one on "The Nuclear Need and its Problems." Pub: Transactions of World Conferences and Regional Meetings; WEO Directory of Energy Information Centres, 2nd ed., 1980; Standard Terms of the Energy Economy, handbook; National Committee special surveys; reports; World Energy Resources 1985-2020, survey (1978); others, list available on request.

World Information Service on Energy (WISE)

See under the Netherlands, p. 89.

# Chapter 6. Nuclear Power in Other Countries of the World

## ARGENTINA

GOVERNMENTAL AGENCY:

Comisión Nacional de Energia Atómica (National Atomic Energy Commission)
Avenida del Libertador 8250
1429 Buenos Aires
Argentina

Founded 1950.

NUCLEAR POWER REACTORS:

Existing:

Atucha-1, near Buenos Aires. Pressurized heavy-water-moderated and cooled reactor, operated by the Comisión Nacional de Energia Atómica. Net power output: 345 MW. Commercial operation since 1974.

Under construction:

Embalse, at Embalse (Cordova). Pressurized heavy-water-moderated and cooled reactor, operated by the Comisión Nacional de Energia Atómica. Net power output: 600 MW. Commercial operation: 1980.

Planned:

Atucha-2, near Buenos Aires. Pressurized heavy-water-moderated and cooled reactor, operated by the Comisión Nacional de Energia Atómica. Net power output: 560 MW. Commercial operation: 1987.

## AUSTRALIA

No plans at present to develop nuclear power (was considered in the 1960s but rejected as uneconomical). Current debate relates to proposed uranium mining and export. Australia has approx. 20 per cent of the known recoverable uranium in non-Communist countries. In 1978, Australia compromised IAEA regulations, terms of the Non-Proliferation Treaty, and rulings of the national Ranger Report (see more below), to secure uranium markets with Finland and the Philippines.

GOVERNMENTAL AGENCIES:

Australian Atomic Energy Commission (AAEC)
45 Beach St.
Coogee, New South Wales 2034
Australia

Composed of five commissioners and one chairperson. Operates under the direction of the Minister for National Resources. Responsible for undertaking and encouraging the search for, and mining of, uranium. Authorized to develop practical uses of atomic energy, and to construct and operate plants. Trains scientists and engineers.

Australia Institute of Nuclear Science and Engineering
Lucas Heights
New South Wales
Australia

Comprised of 17 university programs and the research and training aspects of the Australian Atomic Energy Commission. Awards grants, organizes conferences, arranges for use of AAEC facilities by post-graduate students.

Australia

NONGOVERNMENTAL ORGANIZATIONS:

Anti-uranium mining forces organized after the Ranger Uranium Environmental Inquiry (formed to investigate the government's proposed North Territory mining development) released a report in 1976 pointing out the dangers of uranium mining and recommending public debate on the matter. A second Ranger Report was released May 1977, by which time anti-mining forces known loosely as the Uranium Moratorium, had gathered considerable strength. Uranium mining is opposed on the following points: radioactivity of mill tailings, concern over Australia's responsibility for radioactive waste created in countries importing Australia's uranium, nuclear proliferation, trespass on Aboriginal lands (cases handled by the Northern Land Council), civil liberties (regarding fines and sentences based on the Atomic Energy Act of 1953 and the Approved Defence Projects Protection Act), centralization of control of nuclear power.

Ballarat Environment Centre
Boot Factory
Nolan St.
Ballarat 3350
Australia 053 358 059

Anti-nuclear activities.

Campaign Against Nuclear Energy (CANE)
310 Angas St.
Adelaide 5000
South Australia 08-223 6917

Mainly opposes proposed uranium mining in South Australia, and a uranium enrichment plant proposed by the state government. Also organizes various anti-nuclear activities, including Nuclear Free Zone campaigns. Involved in the Nuclear Free Pacific campaign (which see). Other offices in Perth and Rockhampton. Pub: CANE Newsletter, bi-m.

Campaign Against Nuclear Power (C.A.N.P.)
P.O. Box 238
North Quay
Brisbane, Queensland 4000
Australia 07-221 0188

Resource center. Organizes rallies and anti-nuclear displays; distributes information to schools and libraries; coordinates volunteer activities such as letter-writing and maintaining information stalls; operates as media liaison. Pub: Campaign Against Nuclear Power Newsletter, m. (distribution: 6,000). Other offices in Nambour and Gold Coast.

Christians For Uranium Moratorium
183 Gertrude St.
Fitzroy 3065
Australia 03-419 5588

Community Research Action Centre (CRAC)
Monash University Union
Clayton
Victoria
Australia 3168 03-541 0811, x3125 or 3141

Organized in 1973 to promote research and action by university students in community affairs. Since 1975 has worked in area of nuclear power. Concerned to inform the public about nuclear and uranium industries' activities and governmental policies, and to stimulate public discussion of the nuclear issue. Has developed a visual show, "Uranium and the Alternatives," with the Australian Friends of the Earth (which see), for display in public places. Pub: Uranium Secrets Fallout, a Uranium Decision (analysis of nuclear industry documents CRAC helped FOE to release); discussion papers, reports, leaflets.

Friends of the Earth (FOE)
232 Castlereagh St.
Sydney, New South Wales 2000
Australia 02 235 8037

Anti-nuclear activities. Other offices in Adelaide, South Brisbane, Perth, Canberra City, Collingwood.

Goldfielders Against Nuclear Energy
46 Boundary St.
Kalgoorlie 6430
Western Australia

Movement Against Uranium Mining (MAUM)
P.O. Box K133
Haymarket 2000
Sydney, New South Wales
Australia 02-267 2459

National organization which distributes information on uranium mining to the public and to students; provides speakers and/or materials to schools, community groups, libraries; and serves as a liaison with other Australian anti-nuclear and environmental groups, women's and workers organizations, and international groups. MAUM lobbies the Australian Parliament on a number of different levels: within the Australian Labor Party through liaison with members of the activist group Labor Against Uranium; with trade unions through MAUM's own Trade Union group; at the local council level through MAUM's Nuclear Free Zones campaign. MAUM's radio collective produces "Radio Fallout," a weekly radio program on nuclear-related issues (2SER-FM, 107.5). Pub: MAUM Newsletter, m.; Anti-Uranium Trade Union News, q.

+MAUM
c/o St. Marks Church Hall
George and Moor Sts.
Fitzroy, Victoria 3065
Australia 03-419 1457

+MAUM
88A Myers St.
Geelong, Victoria 3220
Australia 052-419 049

+MAUM
P.O. Box 1875
Canberra City, Australian Capital Territory 2601
Australia 062-473 064

+MAUM
P.O. Box 364, G.P.O.
Townsville, Queensland 4810
Australia 077-716 226

Movement For a Non-Nuclear Future
P.O. Box 2120
Darwin, Northern Territory 5794
Australia 089-813 804

Queensland Trade Union Anti-Nuclear Lobby
P.O. Box 196
Broadway, Brisbane 4000
Australia

Uranium Moratorium
P.O. Box 5017A
Newcastle West, New South Wales 2300
Australia 049 21116

## AUSTRIA

A referendum held November 1978 on the Tullnerfeld nuclear power plant received 50.47 per cent anti-nuclear votes and 49.53 per cent pro-nuclear votes.

### SEMI-GOVERNMENTAL AGENCY:

Oesterreichische Studiengesellschaft fuer Atomenergie Ges. m.b.h. -- SGAE (Austrian Atomic Energy Research Ltd.)
Lenaugasse 10
1082 Vienna
Austria

Founded 1956. Capital shared by the government (51 per cent), state industries (26 per cent), and private enterprise (23 per cent). Pub: SGAE-reports.

### NONGOVERNMENTAL ORGANIZATIONS:

The Austrian anti-nuclear effort has been composed of many small voluntary citizen groups formed to protest specific nuclear power decisions, and later disbanded.

### NUCLEAR POWER REACTORS:

Existing:

Zwentendorf, northwest of Vienna on the Danube. Commercial operation begun 1971, ceased 1978.

Under construction:

Tullnerfeld, in Zwentendorf, Lower Austria. Boiling light-water-cooled and moderated reactor operated by Gemeinschaftskernkraftwerk Tullnerfeld. Net power output: 692 MW. Construction started 1971; all further development halted in 1975 and postponed indefinitely.

## BELGIUM

### GOVERNMENTAL AGENCY:

Commissariat à l'Energie Atomique (Atomic Energy Commission)
Administration de l'Energie
Ministère des Affaires Economiques
rue de Mot 24-26
1040 Brussels
Belgium

Founded 1950.

### NONGOVERNMENTAL ORGANIZATIONS:

Centre d'Etude de l'Energie Nucléaire/Studiecentrum voor Kernenergie -- CEN/SK (Center for the Study of Nuclear Energy)
144 avenue Eugène Plasky
1040 Brussels
Belgium

Founded 1952. A board of industry, science, public administration representatives which aims to train personnel, conduct research, provide experimental facilities for industry. Has three reactors and two critical assemblies at labs in Mol-Doul in northern Belgium.

Institut Interuniversitaire des Sciences Nucléaires (Inter-University Institute of the Nuclear Sciences)
5 rue d'Egmont
1050 Brussels
Belgium

Founded 1947. 150 science researchers. Aims to promote research in advanced teaching establishments.

### NUCLEAR POWER REACTORS:

Existing:

BR-3, at Mol. Pressurized light-water-moderated and cooled reactor, operated by the Groupe Independent pour l'Exploitation du BR-3. Net power output: 10.5 MW. Commercial operation since 1962.

Doel-1, at Doel-Beveren, Flandre. Pressurized light-water-moderated and cooled reactor, operated by the Société Réunie d'Energie de Baffin de l'Escaut SA. Net power output: 392.5 MW. Commercial operation since 1975.

Doel-2, at Doel-Beveren, Flandre. Pressurized light-water-moderated and cooled reactor, operated by the Société Réunie d'Energie de Baffin de l'Escaut SA. Net power output: 392.5 MW. Commercial operation since 1975.

Tihange-1, at Tihange, Liege. Pressurized light-water-moderated and cooled reactor, operated by the Société Intercommunale Belge de Gas et d'Electricité. Net power output: 880 MW. Commercial operation since 1975.

Under construction:

Doel-3, at Doel-Beveren, Flandre. Pressurized light-water-moderated and cooled reactor, operated by the Société Réunie d'Energie de Baffin de l'Escaut SA. Net power output: 897 MW. Commercial operation: 1981.

Tihange-2, at Tihange, Liege. Pressurized light-water-moderated and cooled reactor, operated by the Société Intercommunale Belge de Gas et d'Electricité. Net power output: 902 MW. Commercial operation: 1981.

Planned:

Doel-4, at Doel, Antwerp. Pressurized light-water-moderated and cooled reactor, operated by the Société Réunie d'Energie de Baffin de l'Escaut SA. Net power output: 1006 MW. Commercial operation: 1984.

Tihange-3, at Tihange Liege. Pressurized light-water-moderated and cooled reactor, operated by the Société Intercommunale Belge de Gas et d'Electricité. Net power output: 1006 MW. Commercial operation: 1984.

BOLIVIA

GOVERNMENTAL AGENCY:

Comisión Boliviana de Energia Nuclear (Bolivian Nuclear Energy Commission)
Avenida 6 de Agosto 2905
Casilla 4821
La Paz
Bolivia

Founded 1960. Nuclear engineering research; surveys and exploitation of radioactive materials.

BRAZIL

In 1975, Brazil contracted to buy a complete nuclear fuel cycle from the Federal Republic of Germany. It has made other nuclear arrangement with the Federal Republic as well.

GOVERNMENTAL AGENCIES:

Comissão Nacional de Energia Nuclear (National Nuclear Energy Commission)
Rua General Severiano
90 Botafoga ZC-82
20.000 Rio de Janeiro, RJ
Brazil

Founded 1956.

Empresas Nucleares Brasileiras, S.A. (NUCLE-BRÁS)
Praia do Flamengo 200
22 andar
Rio de Janeiro
Brazil

Founded 1974 to establish fuel cycle industries and promote the transfer of nuclear technology to industries in Brazil. Research and development activities. Has represented Brazil in agreements with four West German firms.

NUCLEAR POWER REACTORS:

Existing:

Angra dos Reis-1, at Rio de Janeiro. Initial operating capacity: 630 MW. Commercial operation: 1979.

Planned:

Angra dos Reis-2 & -3, at Rio de Janeiro. Net power output: 1200 MW each.

BULGARIA

GOVERNMENTAL AGENCY:

Institute of Nuclear Research and Nuclear Energetics of the Bulgarian Academy of Sciences
Lenin Street 72
Sofia
Bulgaria

Founded 1946. Has a research reactor (supplied by the U.S.S.R.) which became operational in 1961 (1500 kw).

NUCLEAR POWER REACTORS:

Existing:

Kozloduy-1, at Kozloduy. Pressurized light-water-moderated and cooled reactor, operated by the State Economic Corporation. Net power output: 408 MW. Commercial operation since 1974.

Kozloduy-2, at Kozloduy. Pressurized light-water-moderated and cooled reactor, operated by the State Economic Coporation. Net power output: 408 MW. Commercial operation since 1975.

Under construction:

Kozloduy-3, at Kozloduy. Pressurized light-water-moderated and cooled reactor, operated by the State Economic Union/Energetics and Coal. Net power output: 408 MW. Commercial operation: 1979.

Kozloduy-4, at Kozloduy. Pressurized light-water-moderated and cooled reactor, operated by the State Economic Union/Energetics and Coal. Net power output: 420 MW. Commercial operation: 1980.

BURMA

GOVERNMENTAL AGENCY:

Union of Burma Atomic Energy Centre
Central Research Organization
Yankin P.O.
Rangoon
Burma

Founded 1955. Performs nuclear research.

CANADA

After India exploded a nuclear device in 1974, Canada imposed a strict ultimatum on its nuclear materials and technology customers: either accept stringent bilateral and multilateral safeguards or have supplies cut off. In January 1977, Canada embargoed all nuclear exports to countries not completing negotiations of revised nuclear safeguard agreements (shipments to Japan, European countries, and the U.S. were affected).

Two of Canada's provincial governments (Saskatchewan and Ontario) have set up inquiry boards to assess the impact of uranium development on their lands. A special group, the Porter Commission, has been set up to study electric power planning, especially Ontario Hydro's commitment to nuclear power.

GOVERNMENTAL AGENCIES:

Atomic Energy Control Board
P.O. Box 1046
Ottawa, Ontario
Canada K1P 5S9

Canada's nuclear regulatory agency.

Atomic Energy of Canada, Ltd.
275 Slater St.
Ottawa, Ontario
Canada

Performs nuclear research and development, and develops and markets nuclear reactors. Designs the CANDU reactors, eight of which are in service in commercial settings; fourteen more are under construction. There is one CANDU unit in India, and one each under construction in the Republic of Korea and Argentina.

Canadian Department of Energy, Mines and Resources
580 Booth St.
Ottawa, Ontario
Canada K1A 0E4 616-995-3065

In August 1977 the Department issued a report and called for a consolidated plan, on radioactive waste. It proposed the following targets: 1978 -- declare a national waste plan and accelerate research and development to effect it; 1983 -- choose at least two hard-rock sites in Ontario for waste disposal; 1985 -- test hard-rock sites; 1988 -- begin building irradiated fuel handling facilities at the sites; 1990 -- start test disposal of immobilized irradiated fuel and wastes; 1995-2000 -- have operating repositories capable of receiving Canada's annual output of irradiated nuclear fuel.

In November 1977, the Department tabled a proposal for new nuclear control and administrative agencies, whereby health, safety, and environmental concerns would be separated from commercial and promotion activities.

NONGOVERNMENTAL ORGANIZATIONS:

The Birch Bark Alliance
c/o Ontario Public Interest Research Group (OPIRG)
Trent University
Peterborough
Ontario, Canada K9J 7B8

Quarterly publication of OPIRG. "Ontario's voice of nuclear concern."

British Columbia Voice of Women (La Voix des Femmes)
104 - 2127 S. Fortieth Ave.
Vancouver, British Columbia
Canada V6M 1W4 604-929-1377

Founded 1960 to protest atomic tests in the atmosphere. Presently engaged in anti-nuclear power and disarmament activities, including letter-writing campaigns, leafletting, film shows, calling for public inquiries, participating in non-violent demonstrations. Pub: The B.C. Voice, q.; pamphlets and fliers.

+Voice of Women
Box 235
NaNaimo, British Columbia
Canada V9R 5J9

Canadian Nuclear Association (CNA)
65 Queen St., West, Ste. 1120
Toronto, Ontario
Canada M5H 2M5 416-363-6433

Founded 1960. Nonprofit voluntary membership organization with the broad purpose of providing "a forum for communication and cooperative action by all groups and individuals interested in meeting the challenges presented by peaceful uses of nuclear energy." CNA accomplishes its goals through a number of standing committees, including Economic Development, Education and Manpower, International Affairs, Government Legislation, Nuclear Safety and Environment, Public Affairs, and Technology. These committees publish reports periodically. CNA also holds an annual International Conference on power reactors, uranium and radioisotopes; sponsors seminars and courses on specific nuclear subjects; holds conferences for science and engineering students; and participates in collective exhibits and papers for trade fairs and conferences. It provides an information services program for its members and the public, and has a small library containing a large collection of nuclear energy reference materials. CNA's technical arm is the Canadian Nuclear Society, which provides "a forum for the exchange of specialized knowledge relating to all aspects of the peaceful uses of nuclear energy." Pub: Nuclear Canada, yearbook; Nuclear Canada, magazine, 11 times/yr.; reports, fliers.

Canadian Coalition for Nuclear Responsibility/Regroupement pour la surveillance du nucléaire (CCNR/Le Regroupement)
2030 MacKay
Montreal, Quebec
Canada H3G 2J1 514-486-6162

Formed 1975. Represents over 200 Canadian citizens groups. Has called on the Canadian government to establish a public inquiry mechanism allowing for "responsible, informed debate on the hazards and benefits of nuclear power development in Canada." Is particularly interested in involving the public while avoiding the spirit of confrontation common to the nuclear debate. CCNR has asked that construction of new nuclear facilities be suspended temporarily, until a safe method of storing radioactive waste is found; in the meantime, it says nuclear construction funds should be channeled into energy conservation, industrial cogeneration, and solar heating in new buildings. Pub: Transitions, q.

+Vancouver Branch, CCNR/Le Regroupement
104 - 2127 W. Fortieth Ave.
Vancouver, British Columbia
Canada V6M 1W4 604-263-7831

Involved primarily in opposing nuclear power and nuclear technology export, and promoting in its stead energy conservation and renewable energy sources. Conducts meetings, hearings, media projects; works to influence Canadian government officials by letter-writing and conducting interviews.. Holds monthly policy formation meetings. Pub: CCNR Newsletter, bi-m.

Carleton University
College of Graduate Studies
Colonel By Drive
Ottawa, Ontario
Canada K1S 5B6 613-231-4403

Research areas include: nuclear reactor safety, uranium research, and nuclear policy.

Energy Probe
43 Queens Park Crescent East
Toronto, Ontario
Canada M5S 2C3 416-978-7014

Project of the Pollution Probe Foundation. Until a safe solution to radioactive waste is found, Energy Probe nuclear policy is: "No Nukes is Good Nukes!" In the meantime, it promotes renewable energy sources, analyzes the economics of various energy systems, and lobbies to pass legislation encouraging conservation and renewable energy. Energy Probe staff are currently participating in activities of the Ontario Energy Board and National Energy Board, and plan to be involved in the Federal Internal Nuclear Review, planned since the national inquiry into nuclear power called for by Canadian citizen groups was cancelled (see p. 73, col. 2). Branch office in Ottawa. Pub: Probe Post, bi-m.; Everything You Wanted to Know About Nuclear Power; Nuclear Energy in Ontario: Who Asked Us?; other publications, complete list available.

Greenpeace
2623 W. Fourth Ave.
Vancouver, British Columbia
Canada V6K 1P8 604-736-0321

Working to bring about a moratorium on uranium mining in Canada -- "the business of supplying the uranium which ends up in nuclear warheads and atomic power plant waste." Mining in Saskatchewan, it is estimated, will supply billions of dollars of uranium in the next ten years to the U.S., Japan, France, and the U.K. Pub: Newsletter, irr.

McGill University
Foster Radiation Laboratory
36 W. University St.
Montreal, PQ
Canada H3A 2B2 514-392-4836

Basic nuclear research.

Maritime Energy Coalition (MEC)
P.O. Box 905, Sta. A
Fredericton, New Brunswick
Canada 506-454-3626

Purpose is to educate the public on the nuclear issue, lobby members of government, and keep the issue "in public view, alive and kicking." Activities include film nights; a speakers bureau; petitioning; organization of pickets, marches and demonstrations in response to current events in government; and research in the areas of waste disposal, alternative energy sources for New Brunswick, and the ethical and social ramifications of nuclear power production and use. Members of MEC have presented briefs to the New Brunswick Emergency Measures Organization regarding a proposed evacuation plan for the Point Lepreau nuclear plant, under construction. MEC is affiliated with the Canadian Coalition for Nuclear Responsibility (which see), and, with CCNR, calls for a national inquiry into nuclear power and the Canadian nuclear policy. Pub: Southern New Brunswick Nuclear News, m.

National Film Board of Canada/Office national du film du Canada (NFB)
P.O. Box 6100
Montreal, Quebec
Canada H3C 3H5

The NFB was formed in 1939 by an Act of the Canadian Parliament; since then it has gained an international reputation for its documentary and animation film. Films available for rental or purchase include several on nuclear power: Douglas Point Nuclear Power Station (description of uranium fission process at Lake Huron site); No Act of God (presentation of various experts' opinions); Nuclear Power (overview); On the Critical Path (how a nuclear pow-

Canada

er plant is constructed); On Power Refuelling (look at efficient refuelling of CANDU reactors); Power from the Atom (electrical production process explained); The Nuclear Age (current account of nuclear research and development in Canada).

Non Nuclear Network (NNN)
121 Avenue Rd.
Toronto, Ontario
Canada M5R 2G3 416-968-3218

Has organized opposition to Ontario Hydro's Darlington complex near Bomanville (Darlington is to be the largest nuclear generating station in the world, with 3400 MW capacity). Opposition is based on safety and environmental concerns, and the arguments that more electricity is unnecessary and nuclear electricity especially uneconomical. NNN has confronted the Atomic Energy Control Board of Canada and Ontario Hydro with regards to freedom of information about nuclear power. It has organized nuclear issues conferences, and a network with anti-nuclear groups in Ontario, Quebec, the rest of Canada, and internationally. Recently, NNN has started a speakers bureau public education campaign, providing speaker training as well as providing speakers and materials to interested groups. Pub: Flyers, reports. NNN events are reported in the Birch Bark Alliance newsletter (see Birch Bark Alliance).

Saskatchewan Coalition Against Nuclear Development
134 Ave. F, South
Saskatoon, Saskatchewan
Canada S7M 1S8 306-652-1571 (day)
306-244-0679 (eve.)

Community organization which aims through public demonstrations and publication activities to raise public concern about nuclear power issues. Program calls for an end to uranium mining in Canada; a local and a provincial referendum on nuclear development; fair settlement of native land claims violated by resource development; full employment through the implementation of renewable energy resources; conservation of existing resources; encouragement of less-developed countries to adopt non-nuclear technologies; the public's right to government information on nuclear power; legislation of a "Green Ban," a workers' boycott of environmentally and socially unacceptable projects. Pub: Crosscurrents, m. (supporting newspaper, available from Sub P.O. #11, Saskatoon, Saskatchewan).

Société pour Vaincre la Pollution (Society to Counter Pollution)
P.O. Box 65
Place d'Armes
Montreal, Quebec
Canada H2Y 3E9 514-844-5477

Research organization which analyzes environmental issues from an ecological perspective. Works to inform and educate the public; also lobbies for ecological solutions to Canadian environmental problems. Where nuclear power is concerned, calls for a moratorium on all nuclear development, and for an open citizen-government-industry debate to discuss the issue and, through objective processes, to reach an equitable, democratic solution to the problem. (The Canadian government had announced that such a debate would take place in 1979, but later cancelled the debate; a Federal Internal Nuclear Review is now planned.) Pub: Journal l'Environment, 6 times/yr.

University of Toronto
Department of Chemical Engineering
Toronto, Ontario
Canada M5S 1A1

Energy researches include development of a model to predict the response of self-powered detectors in reactor radiation environments, and small reactor development.

University of Toronto
Department of Geography
100 St. George St.
Toronto, Ontario
Canada M5S 1A1 416-978-3375

Studies of the environmental impact of various energy projects, and disposal of spent nuclear fuel.

University of Toronto
Department of Mechanical Engineering
Toronto, Ontario
Canada M5S 1A1

Energy research, including application of control techniques for nuclear reactor spatial controls.

University of Toronto
Department of Metallurgy and Materials Science
Toronto, Ontario
Canada M5S 1A1

Works on extraction and melting of nuclear grade metals.

University of Toronto
Institute of Environmental Studies
Toronto, Ontario
Canada M5S 1A1

Study of pollutants using waste modes for heat from nuclear reactors.

NUCLEAR POWER REACTORS:

Existing:

Bruce-1 and -2, at Tiverton, Ontario. Pressurized heavy-water-moderated and cooled reactors, operated by Ontario Hydro. Net power output: 740 MW each. Commercial operation since 1977.

Bruce-3, at Tiverton, Ontario. Pressurized heavy-water-moderated and cooled reactor, operated by Ontario Hydro. Net power output: 740 MW. Commercial operation since 1978.

Existing reactors, cont.:

Bruce-4, at Tiverton, Ontario. Pressurized heavy-water-moderated and cooled reactor, operated by Ontario Hydro. Net power output: 740 MW. Commercial operation since 1979.

Douglas Point Generating Station, at Tiverton, Ontario. Pressurized heavy-water-moderated and cooled reactor, operated by Ontario Hydro. Net power output: 206 MW. Commercial operation since 1968.

Gentilly-1, at Nicolet, Quebec. Heavy-water-moderated, boiling light-water-cooled reactor, operated by Hydro Quebec. Net power output: 250 MW. Commercial operation since 1972.

NPD, at Rolphton, Ontario. Pressurized heavy-water-moderated and cooled reactor, operated by Ontario Hydro. Net power output: 22.5 MW. Commercial operation since 1962.

Pickering-1 and -2, at Pickering, Ontario. Pressurized heavy-water-moderated and cooled reactors, operated by Ontario Hydro. Net power output: 514 MW each. Commercial operation since 1971.

Pickering-3, at Pickering, Ontario. Pressurized heavy-water-moderated and cooled reactor, operated by Ontario Hydro. Net power output: 514 MW. Commercial operation since 1972.

Pickering-4, at Pickering, Ontario. Pressurized heavy-water-moderated and cooled reactor, operated by Ontario Hydro. Net power output: 514 MW. Commercial operation since 1973.

Under construction:

Bruce-5, at Tiverton, Ontario. Pressurized heavy-water-moderated and cooled reactor, operated by Ontario Hydro. Net power output: 750 MW. Commercial operation: 1984.

Bruce-6, at Tiverton, Ontario. Pressurized heavy-water-moderated and cooled reactor, operated by Ontario Hydro. Net power output: 750 MW. Commercial operation: 1983.

Bruce-7, at Tiverton, Ontario. Pressurized heavy-water-moderated and cooled reactor, operated by Ontario Hydro. Net power output: 750 MW. Commercial operation: 1985.

Bruce-8, at Tiverton, Ontario. Pressurized heavy-water-moderated and cooled reactor, operated by Ontario Hydro. Net power output: 750 MW. Commercial operation: 1986.

Darlington-1, at Darlington, Ontario. Pressurized heavy-water-moderated and cooled reactor, operated by Ontario Hydro. Net power output: 854 MW. Commercial operation: 1988.

Darlington-2, at Darlington, Ontario. Pressurized heavy-water-moderated and cooled reactor, operated by Ontario Hydro. Net power output: 854 MW. Commercial operation: 1987.

Darlington-3, at Darlington, Ontario. Pressurized heavy-water-moderated and cooled reactor, operated by Ontario Hydro. Net power output: 854 MW. Commercial operation: 1989.

Darlington-4, at Darlington, Ontario. Pressurized heavy-water-moderated and cooled reactor, operated by Ontario Hydro. Net power output: 854 MW. Commercial operation: 1990.

Gentilly-2, at Nicolet, Quebec. Pressurized heavy-water-moderated and cooled reactor, operated by Hydro Quebec. Net power output: 638 MW. Commercial operation: 1980.

Pickering-5, at Pickering, Ontario. Pressurized heavy-water-moderated and cooled reactor, operated by Ontario Hydro. Net power output: 516 MW. Commercial operation: 1981.

Pickering-6 and -7, at Pickering, Ontario. Pressurized-heavy-water-modified and cooled reactors, operated by Ontario, Hydro. Net power output: 516 MW each. Commercial operation: 1982.

Pickering-8, at Pickering, Ontario. Pressurized heavy-water-moderated and cooled reactor, operated by Ontario Hydro. Net power output: 516 MW. Commercial operation: 1983.

Point Lepreau, at St. John, New Brunswick. Pressurized heavy-water-moderated and cooled reactor, operated by New Brunswick Electric Power Commission. Net power output: 633 MW. Commercial operation: 1981.

## CHILE

GOVERNMENTAL AGENCY:

Comisión Chilena de Energia Nuclear
Los Jesuitas 645
Casílla 188-D
Santiago
Chile

Founded 1965, to develop peaceful uses of nuclear energy. Regulates and controls all nuclear-related matters in Chile.

## CHINA, PEOPLE'S REPUBLIC OF

China has a modest nuclear program, which it is now seeking to expand with U.S. help (it has requested documents from the NRC). Nuclear power development is not a matter of urgent concern for China, since it has abundant fossil fuels; it is being pursued in the interests of minimizing coal pollution and of staying abreast of new technologies.

China does have nuclear arms capacity. It is not party to the Nuclear Proliferation Treaty, or to the Partial Test Ban Treaty of 1963. It is the only nuclear power still to explode nuclear weapons in the atmosphere (none were registered or reported, though, in 1979). China proclaims consistently its adherence to a doctrine of no first-use of nuclear weapons. Recently, it turned down a Libyan request for nuclear weapons.

GOVERNMENTAL AGENCIES:

The Second Ministry of Machine Building is the government branch for China's nuclear power program.

Atomic Energy Institute
Chinese Academy of Sciences
Peking
China

Research activities.

Atomic Energy Research Institute
Shanghai
China

Has a flexible critical reactor assembly modeled on a Westinghouse reactor. Studies of heat transfer in reactor systems.

Atomic Research Centre
Tarim Basin
Sinkiang
China

NUCLEAR POWER REACTORS:

It is believed that China has more than 50 reactors in operation; the exact figure is not available.

CHINA (TAIWAN)

GOVERNMENTAL AGENCIES:

Atomic Energy Council (AEC)
53 Jen Ai Rd.
BCC Building, Sixth Floor
Section 3
Taipei
Taiwan

Pub: Nuclear Science Journal, q.; Chinese AEC Bulletin, every 2 months.

Institute of Nuclear Energy Research (INER)
c/o Atomic Energy Council
P.O. Box 3
Lung Tan
Taiwan 325

National research center.

COLOMBIA

GOVERNMENTAL AGENCY:

Instituto de Asuntos Nuclears -- IAN (Institute of Nuclear Studies)
Avenida Aeropuerto Eldorado
Carrera 50
Apdo. Aéreo 8595
Bogotá
Colombia

Pub: Boletin Bibliográfico de Información.

COSTA RICA

GOVERNMENTAL AGENCY:

Comisión de Energía Atómica de Costa Rica (Atomic Energy Commission of Costa Rica)
Apartado Postal 6681
San José
Costa Rica

CUBA

GOVERNMENTAL AGENCY:

Comisión Nacional para el Uso Pacífico de la Energía Atómica (National Commission for the Peaceful Use of Atomic Energy)
Apartado 2169 9
Havana 2

Founded 1974. Attached to the executive arm of the State Committee for Science and Technology (Comité Estatal de Ciencia y Técnica). Concerned with the peaceful applications of the atom.

NUCLEAR POWER REACTORS:

Under construction:

Cienfuegos-1, at Cienfuegos. Pressurized light-water-moderated and cooled reactor. Net power output: 408 MW. Commercial operation: 1983.

Planned:

Cienfuegos-2, at Cienfuegos. Pressurized light-water-moderated and cooled reactor. Net power output: 408 MW. Commercial operation: 1984.

CZECHOSLOVAKIA

GOVERNMENTAL AGENCIES:

Czechoslovak Atomic Energy Commission (ČSKAE)
Slezská 9
120 29 Prague 2
Czechoslovakia

Responsible for the peaceful utilization of nuclear energy and for coordinating the atomic energy program of Czechoslovakia.

Ministry of Fuel and Power
Vinohradská 8
120 70 Prague 2
Czechoslovakia

Responsible for the construction of nuclear power stations.

Ústar jaderného výzkumu (Institute of Nuclear Research)
250 68 Řež u Prahy
Czechoslovakia

Founded 1955.

NUCLEAR POWER REACTORS:

Existing:

A-1 Bohunice, near Jaslovske Bohunice. Heavy-water-moderated, gas-cooled reactor, operated by Atomova Electraren, Jaslovske Bohunice. Net power output: 110 MW. Commercial operation since 1972.

Bohunice-1, at Trnava, Slovak SR. Pressurized light-water-moderated and cooled reactor, operated by the Slovak Energy Board. Net power output: 380.5 MW. Commercial operation since 1979.

Under construction:

Bohunice-2, at Trnava, Slovak SR. Pressurized light-water-moderated and cooled reactor, operated by the Slovak Energy Board. Net power output: 380.5 MW. Commercial operation: 1980.

Dukovany-1 and -2, in Czech SR. Pressurized light-water-moderated and cooled reactor, operated by the Czechoslovak Energy Board. Net power output: 380.5 MW each. Commercial operation: 1983.

Planned:

Brno, at Brno. Pressurized light-water-moderated and cooled reactor. Net power output: 420 MW.

Dukovany-3 and -4, at Dukovany. Pressurized light-water-moderated and cooled reactors, operated by the Czechoslovak Energy Board. Net power output: 420 MW each.

Levice-1, -2, -3, and -4, in Slovakia. Pressurized light-water-moderated and cooled reactors, operated by the Slovak Energy Board. Net power output: 420 MW each.

## DENMARK

Denmark has no nuclear power plants. It participates in nuclear research conducted under the auspices of CERN (Conseil Européen de la Recherche Nucléaire).

GOVERNMENTAL AGENCIES:

Danish Energy Agency
29 Strandgade
1401 Copenhagen K
Denmark

Founded 1976.

Risø National Laboratory and Atomic Research Station
Forsogsanlaeg Risø
P.O. Box 49
4000 Roskilde
Denmark

Pub: Risø Reports.

NONGOVERNMENTAL ORGANIZATION:

Organisationen til Oplysning om Atomkraft (OOA)
(Organization for Reconsideration of Atomic Energy)
Skindergade 26
DK-1159 Kobenhavn K
Denmark 45-1-110673

Nonpartisan movement which aims to disseminate information about, and therefore informed, critical evaluation of, nuclear power; to increase research evaluation of alternative energy sources; to formulate a long-term far-sighted energy policy which takes into consideration social and ecological factors. OOA's national secretariat, located in Copenhagen, maintains a library of more than 2,000 books, 200 magazines, articles and papers relating to the nuclear power issue. OOA sponsors lectures, hearings, film showings and other informational events. A special task force of OOA covers the Barsebaeck reactors, 20 kilometers from Copenhagen on the Swedish coast, and works for their closure. OOA is a source of information on low-level radiation. Has worked a number of times for a postponement of nuclear power development in Denmark, achieving success. Pub: Atomkraft?, bi-m.; pamphlets, fliers (most in Danish).

## DOMINICAN REPUBLIC

GOVERNMENT

Comisión Nacional de Asuntos Nucleares (National Commission of Nuclear Affairs)
Edificio de la Defensa Civil
Dr. Delgado 58
Santo Domingo, D.N.
Dominican Republic

## ECUADOR

Comisión Ecuatoriana de Energia Atómica (National Atomic Energy Commission)
Cordero 779 y Avenida 6 de Diciembre
Quito
Ecuador

Performs research. Since 1975 has overseen the production and sale of radioactive materials.

## EGYPT

GOVERNMENTAL AGENCIES:

A Higher Nuclear Council, with 32 members and the Prime Minister as the Chair, was created in August 1975.

Atomic Energy Organization
Dokki
Cairo
Egypt

Founded 1955.

NUCLEAR POWER REACTOR:

At Inchasse. Operating capacity: 2000 kw. Became operational: 1961.

EL SALVADOR

GOVERNMENTAL AGENCY:

Comisión Salvadoreña de Energia Nuclear (National Nuclear Energy Commission)
San Salvador
El Salvador

FINLAND

Nuclear energy accounts for just less than 10 per cent of Finland's total electrical energy. Since all nuclear materials and technology are purchased from other ccountries (U.S.S.R., France), foreign politics are a factor in the country's nuclear development.

GOVERNMENTAL AGENCIES:

Atomic Energy Commission
Ministry of Trade and Industry
Rautatielaisenkatu 6
00520 Helsinki 52

Consultative body which advises the Finnish government on nuclear matters. The Ministry of Trade and Industry is the administrative and licensing authority.

NONGOVERNMENTAL ORGANIZATIONS:

Public opinion has not been consulted on the nuclear power issue; only recently have anti-nuclear groups organized, and they are still rather weak. EVY (listed below) is the only organization which is specifically anti-nuclear, though other groups do work in the area.

Energiapoliittinen Yhdistys-Vaihtoehto Ydinvoimalle (EVY) (Energy Policy Organization -- An Alternative to Nuclear Power)
Valpurintie 6
SF-00270 Helsinki 27
Finland 90-417724

Opposed to nuclear power; promotes alternative energy sources. Organizes demonstrations, including one with more than 5,000 participants in Spring 1980. Collects and distributes information related to its work. Pub: Vaihtoehto Ydinvoimalle, q. (English summaries available); membership publications for Finnish and Swedish language members.

Teknikkinenkorkeakoulu (Helsinki University of Technology)
Department of Technical Physics
Otaniemi
Finland

Offers courses and training in nuclear engineering and nuclear physics.

NUCLEAR POWER REACTORS:

Existing:

Loviisa-1, at Loviisa. Pressurized light-water-moderated and cooled reactor, operated by Imatran Voima OY. Net power output: 420 MW. Commercial operation since 1977.

TVO-1, at Olkiluoto, Eurajoki. Boiling light-water-cooled, graphite-moderated reactor, operated by Tellisuuden Voima Oy. Net power output: 660 MW. Commercial operation since 1978.

Under construction:

Loviisa-2, at Loviisa. Pressurized light-water-moderated and cooled reactor, operated by Imatran Voima OY. Net power output: 420 MW. Commercial operation: 1979.

TVO-2, at Olkiluoto, Eurajoki. Boiling light-water-cooled, graphite-moderated reactor, operated by Tellisuuden Voima Oy. Net power output: 660 MW. Commercial operation: 1980.

Planned:

Loviisa-3, at Loviisa. Pressurized light-water-moderated and cooled reactor, operated by Imatran Voima OY. Net power output: 1000 MW. Commercial operation: 1988.

FRANCE

France conducts a fast breeder reactor program. It maintains nuclear fuel reprocessing capacity on a commercial scale at La Hague and Marcoule.

France is one of five major nations to have nuclear weapons capability. For "technical reasons" it is not a party to the Non-Proliferation Treaty of 1968, but it has said it "will conduct itself in this field in the future exactly as the states that decide to sign it" (October 1980). Neither is France a party to the Partial Test Ban Treaty of 1963, but since 1974 it has conducted all its nuclear explosions underground, as the treaty prescribes.

GOVERNMENTAL AGENCIES:

A special council instituted in September 1976 determines France's nuclear foreign policy. This council is chaired by the French President.

Commissariat à l'Energie Atomique (Atomic Energy Commission)
29-33 rue de la Fedération
Paris 15e
France

Founded 1945. Headed by a High Commissioner. Under direct authority of the Ministry of Industry and Research. The Commission is a public corporation with administrative and financial autonomy, and responsibilities in scientific research, technological development and industry in the nuclear field. France's second five-year atomic energy program (1957-1961) arranged for the Commission to cease sole responsibility for nuclear energy in France.

The responsibility is now shared by other corporations (including Electricité de France) which are entrusted with the realization and exploration of nuclear electricity production. The Commission is administered by a 15-member Atomic Energy Committee (Comité de l'Energie Atomique) composed of government officials and representatives from science and industry.

Institut National des Sciences et Techniques Nucléaires (National Institute of Nuclear Sciences and Techniques)
B.P. 6
91190 Gif-sur-Yvette
France

Founded 1956. Affiliated with the Institute are four Nuclear Research Centers (Centres d'Etudes Nucléaires): Cadarache, in Bouches-du-Rhône, founded 1960; Fontenay-aux-Roses, founded 1945; Grenoble, in Grenoble-Cédex, founded 1956; and Saclay, in Gif-sur-Yvette, founded 1949. The Centers are equipped with reactors, and are a means of involving universities in the advancement of nuclear science and of avoiding the over-centralization of research. The Centre at Cadarache is devoted to industrial research, electricity generation and propulsion.

Centre National de la Recherche Scientifique (CNRS) (National Center of Scientific Research)

The Center is comprised of several research laboratories, the largest one of which is at Strasbourg:

Groupe de Laboratoires de Strasbourg-Cronenbourg
rue de Loess
B.P. 20 CRO
67037 Strasbourg Cédex
France

NONGOVERNMENTAL ORGANIZATIONS:

Federation Française des Sociétés de Protection de la Nature (FFSPN) (French Federation of Societies for the Protection of the Environment)
57, rue Cuvier
Paris 5e
France 46 99 59 97

Concerned with a broad range of environmental issues, of which one is the impact of nuclear power on the environment. Pub: Courrier de la Nature, bi-m.

NUCLEAR POWER REACTORS:

Existing:

Ardennes, at Chooz, Ardennes. Pressurized light-water-moderated and cooled reactor, operated by the Société d'Energie Nucléaire Franco-Belge des Ardennes. Net power output: 270 MW. Commercial operation since 1967.

Bugey-1, at Saint-Vulbas Ain. Gas-cooled, graphite-moderated reactor, operated by Electricité de France. Net power output: 540 MW. Commercial operation since 1972.

France

Bugey-2 and -3, at Saint-Vulbas Ain. Pressurized light-water-moderated and cooled reactors, operated by Electricité de France. Net power output: 925 each. Commercial operation since 1978.

Bugey-4, at Saint-Vulbas Ain. Pressurized light-water-moderated and cooled reactor, operated by Electricité de France. Net power output: 905 MW. Commercial operation since 1979.

Chinon-2, at Avoine, Chinon. Gas-cooled, graphite-moderated reactor, operated by Electricité de France. Net power output: 200 MW. Commercial operation since 1966.

Chinon-3, at Avoine, Chinon. Gas-cooled, graphite-moderated reactor, operated by Electricité de France. Net power output: 320 MW. Commercial operation since 1967.

EL-4, at Brennilis, Monts Arrel. Heavy-water-moderated, gas-cooled reactor, operated jointly by the Commissariat à l'Energie Atomique and Electricité de France. Net power output: 70 MW. Commercial operation since 1968.

Fessenheim-1, at Fessenheim, Haut-Rhine. Pressurized light-water-moderated and cooled reactor, operated by Electricité de France. Net power output: 890 MW. Commercial operation since 1977.

Fessenheim-2, at Fessenheim, Haut-Rhine. Pressurized light-water-moderated and cooled reactor, operated by Electricité de France. Net power output: 890 MW. Commercial operation since 1978.

Phenix, at Marcoule, Gard. Fast breeder reactor, operated jointly by the Commissariat a l'Energie Atomique and Electricité de France. Net power output: 250 MW. Commercial operation since 1974.

St. Laurent A1, at St. Laurent des Eaux, Loir et Cher. Gas-cooled, graphite-moderated reactor operated by Electricité de France. Net power output: 480 MW. Commercial operation since 1969.

St. Laurent A2, at St. Laurent des Eaux, Loir et Cher. Gas-cooled, graphite-moderated reactor operated by Electricité de France. Net power output: 515 MW. Commerical operation since 1971.

Under construction:

Blayais-1 and -2, at Braud, Gironde. Pressurized light-water-moderated and cooled reactors, operated by Electricité de France. Net power output: 925 MW each. Commercial operation: 1981.

Bugey-5, at Saint-Vulbas Ain. Pressurized light-water-moderated and cooled reactor, operated by Electricité de France. Net power output: 905 MW. Commercial operation: 1979.

Chinon-B1 and -B2, at Chinon. Pressurized light-water-moderated and cooled reactors, operated by Electricité de France. Net power output: 905 MW each. Commercial operation: 1982.

Dampierre-1, at Dampierre, Loiret. Pressurized

Reactors under construction (cont.):

light-water-moderated and cooled reactor, operated by Electricité de France. Net power output: 905 MW. Commercial operation: 1979.

Dampierre-2 and -3, at Dampierre, Loiret. Pressurized light-water-moderated and cooled reactors, operated by Electricité de France. Net power output: 905 MW each. Commercial operation: 1980.

Dampierre-4, at Dampierre, Loiret. Pressurized light-water-moderated and cooled reactor, operated by Electricité de France. Net power output: 905 MW. Commercial operation: 1981.

Gravelines B1, at Gravelines Nord. Pressurized light-water-moderated and cooled reactor, operated by Electricité de France. Net power output: 925 MW. Commercial operation: 1979.

Gravelines B2, at Gravelines Nord. Pressurized light-water-moderated and cooled reactor, operated by Electricité de France. Net power output: 925 MW. Commercial operation: 1980.

Gravelines B3 and B4, at Gravelines Nord. Pressurized light-water-moderated and cooled reactors, operated by Electricité de France. Net power output: 925 MW each. Commercial operation: 1981.

Paluel-1 and -2, at Seine Maritime. Pressurized light-water-moderated and cooled reactors, operated by Electricité de France. Net power output: 1300 MW each. Commercial operation: 1983.

St. Laurent B1 and B2, at St. Laurent des Eaux, Loir et Cher. Pressurized light-water-moderated and cooled reactors, operated by Electricité de France. Net power output: 905 MW each. Commercial operation: 1981.

Super Phénix, at Creys-Malville. Fast breeder reactor, operated by the Groupement Centrale Nucléaire Européenne de Neutrons Rapides. Net power output: 1200 MW. Commercial operation: 1983.

Tricastin-1 and -2, at Troischateaux Drome. Pressurized light-water-moderated and cooled reactors, operated by Electricité de France. Net power output: 925 MW each. Commercial operation: 1979.

Tricastin-3 and -4, at Troischateaux Drome. Pressurized light-water-moderated and cooled reactors, operated by Electricité de France. Net power output: 925 MW. Commercial operation: 1980.

Planned:

Blayais-3, at Braud, Gironde. Pressurized light-water-moderated and cooled reactor, operated by Electricité de France. Net power output: 925 MW. Commercial operation: 1982.

Blayais-4, at Braud, Gironde. Pressurized light-water-moderated and cooled reactor, operated by Electricité de France. Net power output: 925 MW. Commercial operation: 1983.

Cruas-1 and -2. Pressurized light-water-moderated and cooled reactors, operated by Electricité de France. Net power output: 905 MW. Commercial operation: 1983.

Cruas-3 and -4. Pressurized light-water-moderated and cooled reactors, operated by Electricité de France. Net power output: 905 MW each. Commercial operation: 1984.

Flamanville-1. Pressurized light-water-moderated and cooled reactor, operated by Electricité de France. Net power output: 1300 MW. Commercial operation: 1985.

Flamanville-2. Pressurized light-water-moderated and cooled reactor, operated by Electricité de France. Net power output: 1300 MW. Commercial operation: 1986.

Paluel-3, at Seine Maritime Nord. Pressurized light-water-moderated and cooled reactor, operated by Electricité de France. Net power output: 1300 MW. Commercial operation: 1984.

Paluel-4, at Seine Maritime Nord. Pressurized light-water-moderated and cooled reactor, operated by Electricité de France. Net power output: 1300 MW. Commercial operation: 1985.

St. Maurice-1. Pressurized light-water-moderated and cooled reactor, operated by Electricité de France. Net power output: 1300 MW. Commercial operation: 1985.

St. Maurice-2. Pressurized light-water-moderated and cooled reactor, operated by Electricité de France. Net power output: 1300 MW. Commercial operation: 1986.

## GERMANY, FEDERAL REPUBLIC OF (WEST GERMANY)

In response to anti-nuclear protests, the West German government in 1975 started an information campaign to justify the increased use of nuclear power in the country. West Germany has a fast breeder reactor program. Fuel reprocessing facilities are located in Karlsruhe. Plans for a reprocessing plant at Gorleben, in Lower Saxony, were rejected after an international team of scientists, contracted by the Lower Saxon government, analyzed safety aspects of the proposed facility.

GOVERNMENTAL AGENCY:

Bundesministerium fuer Forschung und Technologie
(Federal Ministry for Research and Technology)
Heinemannstrasse 2
5300 Bonn 2
Postfach 200706
West Germany.

Founded 1955. The Ministry is responsible for promotion of nuclear research and engineering, and for planning and coordinating activities within the framework of the German Energy Research and Energy Technology Program (one three-year phase ended in 1980). It is divded into five departments according to the following functions: administration, planning, finance, coordination of research institutions; coordination of research, and general promotion of research and international cooperation; procurement of energy raw materials, investigation of energy production techniques, research in ecology, medicine; data-processing and information systems; and space travel and transport systems. The Ministry maintains six research centers affiliated with independent institutions, where research is carried out in cooperation with universities and industry. Centers are located in Geesthacht, Karlsruhe, Munich, Berlin, and Juelich.

NONGOVERNMENTAL ORGANIZATIONS:

There has been much protest against nuclear power development in West Germany; it has resulted in a court order halting the construction of three power stations which were scheduled to begin operation in 1981 and 1982.

Bundesverband Burgerinitiatieve Umweltschutz (BBU)
Schiffkopfweg 31a
75 Karlsruhe 21
West Germany 0721-574248

Coalition of West German anti-nuclear groups. Disseminates information, organizes direct actions. Protests proposed nuclear plants, promotes solar energy technologies as alternatives. Pub: Umweltmagazin, bi-m.

Buergerinitiative Umweltschutz Unterelbe (BUU/Hamburg)
Bartelstrasse 26
2000 Hamburg 6
West Germany 439 86 71

Alliance of citizens' organizations in the Hamburg area. Opposed to nuclear power and nuclear weapons. Activities include: demonstrations (including extended effort in opposition to the proposed spent-fuel reprocessing site at Gorleben), information tables, movie shows, theatrical and musical performances, and publishing. Specialized groups within BUU research technological and social aspects of nuclear power; also first aid, and juridical consequences of anti-nuclear activities. Pub: Weg mit dem Atom-Programm (Away With the Atom Program), a short history of BUU; Gorleben Aktuell, journal of the Gorleben protest; fliers, posters, stickers (all in German).

Deutsche Friedensgesellschaft -- Vereinigte Kriegsdienstgegner (DFG-VK) (German Peace Society)
Rellinghauser Str. 214
D-4300 Essen-Sued
West Germany 0201 25999

Pacifist organization dedicated to peace education and action; interested in developing perspectives for international security through collective peaceful coexistence. Pub: Civil Courage, m.

NUCLEAR POWER REACTORS:

Existing:

AVR Juelich, at Juelich. High-temperature gas-cooled, graphite-moderated reactor, operated by Arbeitsgemeinschaft Versuchs-Reaktor GmbH. Net power output: 13.5 MW. Commercial operation since 1969.

Biblis-A, at Biblis Hessen. Pressurized light-water-moderated and cooled reactor, operated by Rheinisch Westfaelisches Elektrizitaetswerk AG. Net power output: 1146 MW. Commercial operation since 1975.

Biblis-B, at Biblis Hessen. Pressurized light-water-moderated and cooled reactor, operated by Rheinisch Westfaelisches Elektrizitaetswerk AG. Net power output 1178 MW. Commercial operation since 1977.

GKN Neckar-1, at Neckar-Westheim, Westheimbaden. Pressurized light-water-moderated and cooled reactor, operated by Kernkraftwerk Neckar. Net power output: 791 MW. Commercial operation since 1976.

KKB Brunsbuettel, at Elbe, Schleswig-Holstein. Boiling light-water-cooled and moderated reactor, operated by Kernkraftwerk Brunsbuettel. Net power output: 770 MW. Commercial operation since 1977.

KKI-Isar-1, at Ohu, Bayern. Boiling light-water-cooled and moderated reactor, operated by Kernkraftwerk Isar. Net power output: 870 MW. Commercial operation since 1979.

KKP Philippsburg-1, at Philippsburg, Baden-West. Boiling light-water-cooled and moderated reactor, operated by Kernkraftwerk Philippsburg. Net power output: 864 MW. Commercial operation since 1979.

KKS Stade, at Stade, Niedersachsen. Pressurized light-water-cooled and moderated reactor, operated by Kernkraftwerk Stade GmbH. Net power output: 630 MW. Commercial operation since 1972.

KKU Unterweser, at Stadland, Niedersachsen. Pressurized light-water-moderated and cooled reactor, operated by Kernkraftwerk Unterweser. Net power output: 1230 MW. Commercial operation: 1979.

KNK II, at Leopoldshafen, Karlsruhe. Fast breeder reactor, operated by Kernkraftwerk-Betriebsgesellschaft mbH. Net power output: 17.8 MW. Commercial operation since 1973.

KRB Gundremmingen-A, at Gundremmingen, Bayern. Boiling light-water-cooled and moderated reactor, operated by Kernkraftwerk RWE-Bayernwerk GmbH. Net power output: 237 MW. Commercial operation since 1967.

KWL Lingen, at Lingen, Niedersachsen. Boiling light-water-cooled and moderated reactor, operated by Kernkraftwerk Lingen GmbH. Net power output: 256 MW. Commercial operation since 1968.

KWO Obrigheim, at Obrigheim, Mosbach. Pressurized light-water-cooled and moderated reactor, operated by Kernkraftwerk Obrigheim GmbH. Net power output: 328 MW. Commercial operation since 1969.

MZFR, at Leopoldshafen, Karlsruhe. Pressurized heavy-water-moderated and cooled reactor, operated by Kernkraftwerk-Betriebsgesellschaft mbH. Net power output: 52 MW. Commercial operation since 1966.

VAK Kahl, at Kahl, Main. Boiling light-water-moderated and cooled reactor, operated by Versuchsatomkraftwerk GmbH. Net power output: 15 MW. Commercial operation since 1962.

Wuergassen, at Wuergassen, Niedersachsen. Boiling light-water-moderated and cooled reactor, operated by Preussische Elektrizitaets AG. Net power output: 640 MW. Commercial operation since 1975.

Under construction:

Brokdorf, at Unterelbe, Schleswig-Holstein. Pressurized light-water-moderated and cooled reactor, operated by Kernkraftwerk Brokdorf GmbH. Net power output: 1288 MW. Commercial operation: 1983.

Grohnde, at Grohnde, Niedersachsen. Pressurized light-water-moderated and cooled reactor, operated by Gemeinschaftskraftwerk Grohnde GmbH. Net power output: 1289 MW. Commercial operation: 1981.

KKG Grafen-Rheinfeld, at Grafen-Rheinfeld, Bayern. Pressurized light-water-moderated and cooled reactor, operated by Bayernwerk AG. Net power output: 1225 MW. Commercial operation: 1980.

KKK Kruemmel, at Kruemmel, Schleswig-Holstein. Boiling light-water-moderated and cooled reactor, operated by Kernkraftwerk Kruemmel. Net power output: 1260 MW. Commercial operation: 1980.

KKP Philippsburg-2, at Philippsburg, Baden-West. Pressurized light-water-moderated and cooled reactor, operated by Kernkraftwerk Philippsburg. Net power output: 1281 MW. Commercial operation: 1983.

KKW Kalkar, at Kalkar. Fast breeder reactor, operated by Gemeinsames Europaeisches Unternehmen. Net power output: 292 MW. Commercial operation: 1983.

KRB Gundremmingen-B and -C, at Gundremmingen, Bayern. Boiling light-water moderated and cooled reactors, operated by Kernkraftwerk RWE-Bayernwerk GmbH. Net power output: 1244 MW each. Commercial operation: 1982.

Muelheim-Kaerlich, at Muelheim-Kaerlich, Rheinlandpfalz. Pressurized light-water-moderated and cooled reactor, operated by Rheinisch-Westfaelisches Elektrizitaetswerk AG. Net power output: 1215 MW. Commercial operation: 1981.

THTR-300, at Hamm-Uentrop. High-temperature gas-cooled, graphite-moderated reactor, operated by Hochtemperatur-Kernkraftwerk GmbH. Net power output: 300 MW. Commercial operation: 1981.

Planned:

Biblis-C, at Biblis, Hessen. Pressurized light-water-moderated and cooled reactor, operated by Rheinisch-Westfaelisches Elektrizitaetswerk AG. Net power output: 1166 MW. Commercial operation: 1983.

Emsland, at Lingen. Pressurized light-water-moderated and cooled reactor, operated by KLE. Net power output: 1232 MW. Commercial operation: 1986.

GKN Neckar-2, at Neckar-Westheim, Westheimbaden. Pressurized light-water-moderated and cooled reactor, operated by Kernkraftwerk Neckar. Net power output: 750 MW. Commercial operation: 1986.

KKH Hamm, at Hamm. Pressurized light-water-moderated and cooled reactor, operated by KLE. Net power output: 1231 MW. Commercial operation: 1985.

KKI Isar-2, at Ohu, Bayern. Operated by Kernkraftwerk Isar. Net power output: 1230 MW. Commercial operation: 1986.

KWB Borken, at Borken, Hessen. Pressurized light-water-moderated and cooled reactor, operated by Preussenelektra. Net power output: 1200 MW.

KWS Wyhl-1, at Emmendingen, Baden-West. Pressurized light-water-moderated and cooled reactor, operated by Kernkraftwerk Sued. Net power output: 1284 MW. Commercial operation: 1986.

Neupotz-1 and -2, at Neupotz, Rheinlandpfalz. Pressurized light-water-moderated and cooled reactors, operated by Rheinisch-Westfaelisches Elektrizitaetswerk AG. Net power output: 1245 MW each. Commercial operation: 1985.

Rehling. Light-water-cooled and moderated reactor. Net power output: 1300 MW.

SNR-2, at Kalkar. Fast breeder reactor, operated by ESK. Net power output: 1300 MW. Commercial operation: 1989.

Vahnum-1 and -2, at Vahnum. Pressurized-light-water-moderated and cooled reactors, operated by Rheinisch-Westfaelisches Elektrizitaetswerk AG. Net power output: 1232 MW each.

## GERMANY, DEMOCRATIC REPUBLIC (EAST GERMANY)

### GOVERNMENTAL AGENCIES:

Staatliches Amt fuer Atomsicherheit und Strahlenschutz der D.D.R. (Board of Nuclear Safety and Radiation Protection of the G.D.R.)
Waldowallee 117
1157 Berlin-Karlshort
East Germany

East Germany agency for nuclear legislation and licensing, radiation protection, and environmental concerns with respect to nuclear power development (especially radioactive waste processing and disposal). Concerned also with theoretical research.

Ministerium fuer Wissenschaft und Technik (Ministry of Science and Technology)
Koepenickerstr. 80/82
102 Berlin
East Germany

Research and development of nuclear power.

Zentralinstitut fuer Kernforschung der AdW (Central Institute for Nuclear Research)
Postfach 19
8051 Dresden
East Germany

NUCLEAR POWER REACTORS:

Operating:

Bruno Leuschner-1 and -2, at Greifswald, Lubmin. Pressurized light-water-moderated and cooled reactors, operated by VEB KKW Bruno Leuschner. Net power output: 408 MW each. Commercial operation since 1974.

Bruno Leuschner-3, at Greifswald, Lubmin. Pressurized light-water-moderated and cooled reactor, operated by VEB KKW Bruno Leuschner. Net power output: 408 MW. Commercial operation since 1978.

Rheinsberg AKW, at Gransee, Rheinsberg. Pressurized light-water-moderated and cooled reactor, operated by VEB KKW Bruno Leuschner. Net power output: 62.5 MW. Commercial operation since 1966.

Under construction:

Bruno Leuschner-4, at Greifswald, Lubmin. Pressurized light-water-moderated and cooled reactor, operated by VEB KKW Bruno Leuschner. Net power output: 408 MW. Commercial operation: 1980.

Magdeburg-1, -2, -3 and -4. Pressurized light-water-moderated and cooled reactors. Net power output: 408 MW each. Commercial operation: 1980.

Planned:

Magdeburg-5, -6, -7 and -8. Pressurized light-water-moderated and cooled reactors. Net power output: 408 MW each.

## GREECE

GOVERNMENTAL AGENCY:

Elliniki Epitropi Atomikis Energias (Greek Atomic Energy Commission)
Aghia Paraskevi
Attikis, Athens
Greece

Founded 1954. Five-member administrative committee.

Democritos Nuclear Research Centre
Aghia Paraskevi
Attiki, Athens
Greece

Nuclear research laboratories; includes a radioactive waste treatment installation.

## GUATEMALA

GOVERNMENTAL AGENCY:

Instituto Nacional de Energia Nuclear (National Institute of Nuclear Energy)
3A Avenida "A" 2-68
Zona 1, Apdo. 1421
Guatemala City
Guatemala

Research into the application of nuclear energy in industry.

## HONDURAS

GOVERNMENTAL AGENCY:

Comisión Hondureña de Energia Atómica (National Atomic Energy Commission)
Apartado 104
Tegucigalpa
D.C.
Honduras

## HUNGARY

Hungary is a member of the Joint Institute for Nuclear Research in Dubna, near Moscow. It has signed agreements for cooperation in the peaceful use of nuclear energy with Czechoslovakia, the German Democratic Republic (East Germany), India, Poland, Romania, and the U.S.S.R.

A Hungarian research reactor at Csillebérc, near Budapest, inaugurated in 1959, was supplied by the U.S.S.R.

GOVERNMENTAL AGENCIES:

Országos Atomenergia Bizottság (National Atomic Energy Commission)
P.O. Box 565
H-1374 Budapest 5
Hungary

Magyar Tudományos Akadémia Atommag Kutato Intézete (Nuclear Research Institute of the Hungarian Academy of Sciences)
Bem ter 18/c
H-4026 Debrecen
Hungary

Founded 1954. Nuclear research. Pub: ATOMKI Közlemények (Bulletin ATOMKI).

NONGOVERNMENTAL ORGANIZATION:

Technical University of Budapest
Budapest
Hungary

Has a low-power training reactor of Hungarian design which became operational in 1970; it is used for university and post-graduate training and research.

NUCLEAR POWER REACTORS:

Under construction:

Paks-1, at Paks, Tolna. Pressurized light-water-moderated and cooled reactor, operated by the Hungarian Electric Works. Net power output: 408 MW. Commercial output: 1980.

Paks-2, at Paks, Tolna. Pressurized light-water-moderated and cooled reactor, operated by the Hungarian Electric Works. Net power output: 408 MW. Commercial operation: 1983.

Planned:

Paks-3, at Paks, Tolna. Pressurized light-water-moderated and cooled reactor, operated by the Hungarian Electric Works. Net power output: 408 MW. Commercial operation: 1984.

Paks-4, Paks, Tolna. Pressurized light-water-moderated and cooled reactor, operated by the Hungarian Electric Works. Net power output: 408 MW. Commercial operation: 1985.

## INDIA

India has fuel reprocessing plants on a commercial scale at Trombay and Tarapur.

In May 1974, India exploded a nuclear weapons device, causing a good deal of international concern. Since then, India has maintained a policy of nonproduction of nuclear weapons, but it has stated it may reconsider if Pakistan, its neighbor and traditional enemy, continues its attempts to produce nuclear weapons.

GOVERNMENTAL AGENCIES:

Atomic Energy Commission
Chhatrapati Shivaji Maharaj Marg.
Bombay 400039
India

Organizes researches into peaceful uses of nuclear energy.

Bhabha Atomic Research Centre (BARC)
Trombay
Bombay 400085
India

Founded 1957. National center for research and development of nuclear energy for peaceful uses. The Centre was responsible for exploding India's first underground nuclear device in 1974 at Pokran, Rajasthan State.

NUCLEAR POWER REACTORS:

Operating:

Tarapur-1 and -2, at Boisar, Maharastra. Boiling light-water-cooled and moderated reactors, operated by the Department of Atomic Energy. Net power output: 198 MW. Commercial operation since 1969.

Under construction:

Kalpakkam-1, at Kalpakkam, Tamilnadu. Pressurized heavy-water-moderated and cooled reactor, operated by the Department of Atomic Energy. Net power output: 220 MW. Commercial operation: 1981.

Kalpakkam-2, at Kalpakkam, Tamilnadu. Pressurized heavy-water-moderated and cooled reactor, operated by the Department of Atomic Energy. Net power output: 220 MW. Commercial operation: 1983.

Narora-1, at Narora, Uttar Pradesh. Pressurized heavy-water-moderated and cooled reactor, operated by the Department of Atomic Energy. Net power output: 220 MW. Commercial operation: 1984.

Narora-2, at Narora, Uttar Pradesh. Pressurized heavy-water-moderated and cooled reactor, operated by the Department of Atomic Energy. Net power output: 220 MW. Commercial operation: 1985.

Rajasthan-2, at Kota, Rajasthan. Pressurized heavy-water-moderated and cooled reactor, operated by the Department of Atomic Energy. Net power output: 207 MW. Commercial operation: 1980.

## INDONESIA

GOVERNMENTAL AGENCY:

Badan Tenaga Atom Nasional (National Atomic Energy Agency)
Jalan Falatehan 1/26
Blok-K. V.
Kebayoran Baru
Jakarta Selatan
Indonesia

Established 1958. Oversees some nuclear research (a research reactor was installed in Indonesia in 1964). Pub: Majalah Batan (Atom Indonesia).

## IRAN

The late Shah of Iran, while he was in power, decided to develop nuclear power in the country. His Eighth Development Plan (1988-93) projected nuclear power accounting for 52 per cent of Iran's total energy. With the Revolution, the future of nuclear development is uncertain.

GOVERNMENTAL AGENCIES:
(Existing before the Revolution)

Atomic Energy Organization of Iran
P.O. Box 12-1198
Teheran
Iran

Established 1973, to produce nuclear power to provide for Iran's base electrical needs, and to secure fuel needs of the nuclear energy program; also to promote use of nuclear energy in industry and medicine, and to provide research and development work and training to develop greater national self-sufficiency in nuclear technology.

Teheran University Nuclear Centre
Institute of Nuclear Science and Technology
P.O. Box 2989
Teheran
Iran

Founded 1958. Performs research, training, and advises on the peaceful applications of nuclear energy. Has two research reactors.

NUCLEAR POWER REACTORS:

Under construction (prior to the Revolution):

Iran-1, at Halileh, Bushehr. Pressurized light-water-moderated and cooled reactor, operated by the Atomic Energy Organization of Iran. Net power output: 1200 MW. Commercial operation: 1980.

Iran-2, at Halileh, Bushehr. Pressurized light-water-moderated and cooled reactor, operated by the Atomic Energy Organization of Iran. Net power output: 1200 MW. Commercial operation: 1981.

Iran-3, at Darkhowin, Khuzestan Province. Pressurized light-water-moderated and cooled reactor, operated by the Atomic Energy Organization of Iran. Net power output: 891 MW. Commercial operation: 1983.

Iran-4, at Darkhowin, Khuzestan Province. Pressurized light-water-moderated and cooled reactor, operated by the Atomic Energy Organization of Iran. Net power output: 991 MW. Commercial operation: 1984.

Planned (prior to the Revolution):

Iran-5 and -6, at Isfahan. Pressurized light-water-moderated and cooled reactors. Net power output: 1200 MW each.

Iran-7 and -8, at Saveh, Central Province. Pressurized light-water-moderated and cooled reactors. Net power output: 1200 MW each.

## IRAQ

Iraq has discussed a nuclear power trade with Brazil.

GOVERNMENTAL AGENCY:

Atomic Energy Commission
Tuwaitha
P.O. Box 765
Baghdad
Iraq

Founded 1956.

Nuclear Research Institute
Baghdad
Iraq

Associated with the Atomic Energy Commission.

## ISRAEL

Israel has no well-defined nuclear power policy at this date. Some scientists and government officials promote nuclear as a dependable supply of electrical energy to sustain the country's growing GNP. Problems of nuclear development for Israel include: Acquisition depends on U.S. willingness to sell reactors (Israel cannot or will not do business with other countries), which in turn depends on relative stability in the Middle East; the cost could run at least half the present annual budget ($5-6 billion of $10-12 billion), and efficiency of nuclear plants (58 per cent or less) could further limit their cost-effectiveness; locating a steady uranium supply; Israel's small size (one-sixth that of the state of Pennsylvania) means that if an accident occurred and radiation were released, the whole country could receive lethal or very high doses of radiation; Israel's small size also makes waste storage a big problem; and finally, nuclear materials could become the object of terrorism or enemy attack.

It is not known whether Israel has nuclear weapons capability. U.S. Central Intelligence Agency reports in the past suggest that it does.

GOVERNMENTAL AGENCY:

Israel Atomic Energy Commission
26 Rehov HaUniversita
P.O. Box 17120
Ramat Aviv
Tel-Aviv
Israel

Established 1952. Chaired by the Prime Minister. Advises the government on long-term policies and priorities in the advancement of nuclear research and development. Supervises implementation of government-approved policies, including licensing of nuclear power plants. Represents Israel in relations with scientific institutions abroad and in international organizations concerned with nuclear matters. Has two research and development centers: Nahal Soreq, in Yavne, and Negev, in Dimona.

NONGOVERNMENTAL ORGANIZATIONS:

Opposition to nuclear power has been organized only recently, and maintains a low profile. Listed below are two research institutes involved in theoretical and practical nuclear research.

Weizmann Institute of Science
Department of Nuclear Physics
Rehovot
Israel

Technion: Israel Institute of Technology
Department of Nuclear Engineering
Haifa
Israel

NUCLEAR POWER REACTORS:

Planned:

Israel-1. Net power output: 600 MW. Commercial operation: 1985.

ITALY

GOVERNMENTAL AGENCIES:

Comitato Nazionale per l'Energia Nucleare - CNEN
Via Regina Margherita 125
00198 Rome
Italy

Established 1960. Supervises pure and applied research; responsible for maintaining technical control over industry and nuclear power plants; promotes development of nuclear energy in industry; cooperates on an international level in nuclear matters. Pub: Notiziano, m.

Ente Nazionale per l'Energia Elettrica (ENEL)
Via G.B. Martini (Piazza Verdi)
Rome
Italy

The Italian State Power Agency. Operates several nuclear power stations (see below).

NUCLEAR POWER REACTORS:

Operating:

Caorso, at Caorso, Piacenza. Boiling light-water-moderated and cooled reactor, operated by ENEL (see above). Net power output: 840 MW. Commercial operation since 1978.

Garigliano, at Sessa Aurunea, Caserta. Boiling light-water-moderated and cooled reactor, operated by ENEL. Net power output: 150 MW. Commercial operation since 1964.

Latina, at Borgo Sabotino Latina. Gas-cooled, graphite-moderated reactor, operated by ENEL. Net power output: 150 MW. Commercial operation since 1964.

Trino Vercellese, at Trino Vercellese, West Vercelli. Pressurized light-water-cooled and moderated reactor, operated by ENEL. Net power output: 242 MW. Commercial operation since 1965.

Under construction:

Cirene, at Cirene, Latina. Heavy-water-moderated, boiling light-water-cooled reactor, operated by ENEL. Net power output: 36 MW. Commercial operation: 1982.

ENEL-6, at Tarquinia, Montaldo Di Castro. Boiling light-water-moderated and cooled reactor, operated by ENEL. Net power output: 980 MW. Commercial operation: 1983.

ENEL-8, at Tarquinia, Montado Di Castro. Boiling light-water-moderated and cooled reactor, operated by ENEL. Net power output: 980 MW. Commercial operation: 1984.

Planned:

ENEL-5 and -7, at Compobasso Vasto. Pressurized light-water-moderated and cooled reactors, operated by ENEL. Net power output: 950 MW. Commercial operation: 1984.

JAPAN

Japan has a fast breeder reactor program. It has nuclear fuel reprocessing facilities at Tokai Mura.

GOVERNMENTAL AGENCIES:

Japan Atomic Energy Commission (JAEC)
2-2-1 Kasumigaseki
Chiyoda-ku
Tokyo
Japan

Policy board for research and development of peaceful uses of nuclear energy. Founded 1955.

Japan Nuclear Safety Commission (JNSC)
2-2-1 Kasumigaseki
Chiyoda-ku
Tokyo
Japan

Established 1978. Responsible for all safety regulation matters.

Atomic Energy Bureau (AEB)
2-2-1 Kasumigaseki
Chiyoda-ku
Tokyo
Japan

Encourages research and development in nuclear energy.

NONGOVERNMENTAL ORGANIZATIONS:

Japan Atomic Industrial Forum (JAIF)
1-1-13 Shinbashi
Minato-ku
Tokyo
Japan

Concern for private industry activities as they relate to the peaceful uses of atomic energy.

Japan's main anti-nuclear movement is known broadly as the Movement To Ban Nuclear Weapons. It is not the work of voluntary citizens' groups but of coalitions of trade unions and political parties. Three of these groups are listed first below.

Gensuikin (Japan Congress Against A- and H-Bombs)
Akimoto Building, Fourth Floor
2019 Tsukasa-cho
Kana, Chiyoda-ku, Tokyo
Japan (Tokyo) 03-294-3994

The largest anti-nuclear coalition in Japan, with nearly 3 million members. Includes three labor associations

(Sohyo, which concentrates on forming labor and political fronts; SHIN SAMBETSU; and the main unions of CHURITSU ROREN), and no political parties. Members of the following parties do cooperate with Gensuikin, however: Japan Socialist Party, Clean Government Party, League of Social Democracy, Japan Communist Party.

Gensuikin works for a nuclear-free world. It is opposed both to military and to commercial, "peaceful" applications of nuclear power; in fact it claims to make no distinction between the two, "on the ground that both are destroying the nation's health." Distributes information, holds meetings and demonstrations, performs research. Organizes an annual Atomic Bomb Disaster Anniversary Conference in Hiroshima and Nagasaki. Helps to organize the annual World Conference Against Atomic and Hydrogen Bombs. Pub: Gensuikin News, m. (Japanese edition), and q. (English edition).

Gensuikyo (Japan Council Against A- and H-Bombs)
c/o Japan Communist Party
Tokyo
Japan

An organization of the Communist Party in Japan; includes members of the Japan Peace Committee. Total membership is approx. 444,000. Gensuikyo is generally opposed by trade unionists. It calls for "an absolute ban on nuclear weapons, and relief for the survivors (of nuclear attacks, called 'Hibakusha')."

Kakukin Kaigi (Congress to Ban Nuclear Weapons)
Hiroshima
Japan

Strong association with over two million members. Controlled by DOMEI, a labor association joined with the Social Democratic Party of Japan. DOMEI has traditionally worked to cooperate with employers.

NAMAZU Collective
2-12-2, Asahimachi
Abeno, Asaka
Japan 06-637-4089

Dual focus on nuclear power and feminism: "the critique of nuclear power deserves an integral place in the attack on male society"; nuclear power is "the climax of the urge towards centralization and alienation that feminism attacks." Involved in demonstrations, leafletting, anti-militarism, and anti-prison activities. Pub: NAMAZU (in English), irr.; Japanese language newsletters.

The Ohdake Foundation
Central Building, Ninth Floor
1-1-5 Kyobashi, Chuo-ku, Tokyo
104 Japan (Tokyo) 272-3900

Research foundation. Distributes information on energy and environmental matters. Pub: Revealing Japan, a monthly collection of news clippings from Japanese and foreign newspapers on energy, environment and social issues; large section on nuclear energy.

People's Research Institute on Energy and Environment
B. Kaikan, Shinjuku 7-26-24
Shinjuku-ku, Tokyo
Japan (160) 03-202-8031

Politically independent study group established 1978. Diverse membership of specialists and interested citizens. Serves to inform the public of anti-nuclear movement activities in Japan. Pub: PRIEE News (Shiryō Geppō), bi-m.

NUCLEAR POWER REACTORS:

Operating:

Fugen ATR, at Myoin, Fukui. Heavy-water-moderated boiling light-water-cooled reactor, operated by the Power Reactor & Nuclear Fuel Development Corporation. Net power output: 150 MW. Commercial operation since 1979.

Fukushima I-1, at Okuma, Fukushima. Boiling light-water-moderated and cooled reactor, operated by Tokyo Electric Power Company. Net power output: 439 MW. Commercial operation since 1971.

Fukushima I-2, at Okuma, Fukushima. Boiling light-water-moderated and cooled reactor, operated by Tokyo Electric Power Company. Net power output: 760 MW. Commercial operation since 1974.

Fukushima I-3, at Okuma, Fukushima. Boiling light-water-moderated and cooled reactor, operated by Tokyo Electric Power Company. Net power output: 760 MW. Commercial operation since 1976.

Fukushima I-4 and I-5, at Okuma and Futaba, Fukushima, respectively. Boiling light water-moderated and cooled reactors, operated by Tokyo Electric Power Company. Net power output: 760 MW each. Commercial operation since 1978.

Genkai-1, at Genkai, Saga. Pressurized light-water-moderated and cooled reactor, operated by Kyushu Electric Power Company. Net power output: 529 MW. Commercial operation since 1975.

Hamaoka-1, at Hamaoka, Shizuoka. Boiling light-water-moderated and cooled reactor, operated by Chubu Electric Power Company. Net power output: 515 MW. Commercial operation since 1976.

Hamaoka-2, at Hamaoka, Shizuoka. Boiling light-water-moderated and cooled reactor, operated by Chubu Electric Power Company. Net power output: 814 MW. Commercial operation since 1978.

Operating reactors, cont.:

Ikata-1, at Ikata, Ehime. Pressurized light-water-moderated and cooled reactor, operated by Shikoku Electric Power Company. Net power output: 538 MW. Commercial operation since 1977.

JPDR-II, at Tokaimura, Ibaraki. Boiling light-water-moderated and cooled reactor, operated by Japan Atomic Energy Research Institute. Net power output: 10.3 MW. Commercial operation since 1963.

Mihama-1, at Mihama, Fukui. Pressurized light-water-moderated and cooled reactor, operated by Kansai Electric Power Company. Net power output: 320 MW. Commercial operation since 1970.

Mihama-2, at Mihama, Fukui. Pressurized light-water-moderated and cooled reactor, operated by Kansai Electric Power Company. Net power output: 470 MW. Commercial operation since 1972.

Mihama-3, at Mihama, Fukui. Pressurized light-water-moderated and cooled reactor, operated by Kansai Electric Power Company. Net power output: 780 MW. Commercial operation since 1976.

Ohi-1 and -2, at Ohi, Fukui. Pressurized light-water-moderated and cooled reactors, operated by Kansai Electric Power Company. Net power output: 1120 MW each. Commercial operation since 1979.

Shimane, at Kashima, Shimane. Boiling light-water-moderated and cooled reactor, operated by Chugoku Electric Power Company. Net power output: 439 MW. Commercial operation since 1974.

Takahama-1, at Takahama, Fukui. Pressurized light-water-moderated and cooled reactor, operated by Kansai Electric Power Company. Net power output: 780 MW. Commercial operation since 1974.

Takahama-2, at Takahama, Fukui. Pressurized light-water-moderated and cooled reactor, operated by Kansai Electric Power Company. Net power output: 780 MW. Commercial operation since 1975.

Tokai-1, at Tokaimura, Ibaraki. Gas-cooled, graphite-moderated reactor, operated by Japan Atomic Power Company. Net power output: 158.5 MW. Commercial operation since 1966.

Tokai-2, at Tokaimura, Ibaraki. Boiling light-water-moderated and cooled reactor, operated by Japan Atomic Energy Company. Net power output: 1056 MW. Commercial operation since 1978.

Tsuruga, at Tsuruga, Fukui. Boiling light-water-moderated and cooled reactor, operated by Japan Atomic Power Company. Net power output: 340 MW. Commercial operation since 1970.

Under construction:

Fukushima I-6, at Futaba, Fukushima. Boiling light-water-moderated and cooled reactor, operated by Tokyo Electric Power Company. Net power output: 1067 MW. Commercial operation: 1979.

Fukushima II-1, at Naraha, Fukushima. Boiling light-water-moderated and cooled reactor, operated by Tokyo Electric Power Company. Net power output: 1067 MW. Commercial operation: 1982.

Genkai-2, at Genkai, Saga. Pressurized light-water-moderated and cooled reactor, operated by Kyushu Electric Power Company. Net power output: 529 MW. Commercial operation: 1981.

Ikata-2, at Ikata, Ehime. Pressurized light-water-moderated and cooled reactor, operated by Shikoku Electric Power Company. Net power output: 538 MW. Commercial operation: 1982.

Kashiwazaki-1, at Kariwa, Niigata. Boiling light-water-moderated and cooled reactor, operated by Tokyo Electric Power Company. Net power output: 1067 MW. Commercial operation: 1984.

Sendai-1, at Sendai, Kagoshima. Pressurized light-water-moderated and cooled reactor, operated by Kyushu Electric Power Company. Net power output: 846 MW. Commercial operation: 1984.

Planned:

Fukushima II-2, at Naraha, Fukushima. Boiling light-water-moderated and cooled reactor, operated by Tokyo Electric Power Company. Net power output: 1067 MW. Commercial operation: 1983.

Fukushima II-3, at Naraha, Fukushima. Boiling light-water-moderated and cooled reactor, operated by Tokyo Electric Power Company. Net power output: 1067 MW. Commercial operation: 1984.

Fukushima II-4, at Futaba, Fukushima. Boiling light-water-moderated and cooled reactor, operated by Tokyo Electric Power Company. Net power output: 1067 MW. Commercial operation: 1985.

Hamaoka-3, at Hamaoka, Shizuoka. Boiling light-water-moderated and cooled reactor, operated by Chubu Electric Power Company. Net power output: 1050 MW. Commercial operation: 1985.

Monju. Liquid metal fast breeder reactor, operated by Power Reactor & Nuclear Fuel Development Corporation. Net power output: 250 MW. Commercial operation: 1987.

Onagawa-1, at Onagawa, Miyagi. Boiling light-water-moderated and cooled reactor, operated by Tohoku Electric Power Company. Net power output: 500 MW. Commercial operation: 1983.

Sendai-2, at Sendai, Kagoshima. Pressurized light-water-moderated and cooled reactor, operated by Kyushu Electric Power Company. Net power output: 846 MW. Commercial operation: 1985.

Takahama-3, at Takahama, Fukui. Pressurized light-water-moderated and cooled reactor, operated by Kansai Electric Power Company. Net power output: 830 MW.

Planned reactors (cont.):

Commercial operation: 1983.

Takahama-4, at Takahama, Fukui. Pressurized light-water-moderated and cooled reactor, operated by Kansai Electric Power Company. Net power output: 830 MW. Commercial operation: 1984.

Tsuruga-2, at Tsuruga, Fukui. Pressurized light-water-moderated and cooled reactor, operated by Japan Atomic Power Company. Net power output: 1115 MW. Commercial operation: 1986.

## KOREA, REPUBLIC OF (SOUTH KOREA)

### GOVERNMENTAL AGENCIES:

Atomic Energy Commission
Ministry of Science and Technology
Seoul 110
Korea

Responsible for basic planning and policy, and furtherance of research and training of nuclear personnel.

Atomic Energy Bureau
Ministry of Science and Technology
Seoul 110
Korea

Administrative body, founded in 1973. Consists of four divisions: Planning, Radiation Safety, Nuclear Reactor Regulation and Licensing, and Nuclear Reactor Technology. The Bureau includes the National System of Safeguards and International Co-operation Directorate.

Korea Atomic Energy Research Institute
P.O. Box 7
Cheong Ryang
Seoul
Korea

Responsible for the localization of nuclear power technology, nuclear manpower development, industrial and medical uses of radiation, and environmental research. Has two reactors. Founded 1973.

### NUCLEAR POWER REACTORS:

Operating:

Ko-Ri-1, at Ko-Ri, Kyongsangnam. Pressurized light-water-moderated and cooled reactor, operated by Korea Electric Company. Net power output: 564 MW. Commercial operation since 1978.

Under construction:

Ko-Ri-2, at Ko-Ri, Kyongsangnam. Pressurized light-water-moderated and cooled reactor, operated by Korea Electric Company. Net power output: 605 MW. Commercial operation: 1983.

Wolsung-1, at Naah-Ri, Kyongsangnam. Pressurized heavy-water-moderated and cooled reactor, operated by Korea Electric Company. Net power output: 628.6 MW. Commercial operation: 1982.

Planned:

Ko-Ri-3, at Ko-Ri, Kyongsangnam. Pressurized light-water-moderated and cooled reactor, operated by Korea Electric Company. Net power output: 930 MW. Commercial operation: 1985.

Ko-Ri-4, at Ko-Ri, Kyongsangnam. Pressurized light-water-moderated and cooled reactor, operated by Korea Electric Company. Net power output: 930 MW. Commercial operation: 1986.

## LUXEMBOURG

### GOVERNMENTAL AGENCY:

Conseil National de l'Energie Nucléaire - CNEN (National Nuclear Energy Council)
Minister of Power
19 blvd. Royal
Luxembourg-Ville
Luxembourg

Chaired by Luxembourg's Minister of Power. Studies economic, legal, financial and technical aspects of nuclear energy use, especially in industrial settings. Takes part in the work of comparable foreign bodies.

## LIBYA

Libya signed the Nuclear Non-Proliferation Treaty in 1975.

In 1976, Libya appealed to the U.S. for nuclear technology to be used for energy and agricultural purposes. Also in 1976, France agreed to provide Libya with a 600 MW power plant, to be built by the French and operated in conjunction with a French water-desalination plant. This nuclear plant would not be able to produce nuclear weapons materials.

In October 1978, the U.S.S.R. announced that it had signed an accord with Libya, to build a complete nuclear power complex. This would make Libya the only non-satellite country, besides Finland, to which the U.S.S.R. has sold nuclear materials and technology.

## MEXICO

### GOVERNMENTAL AGENCIES:

Comisión Nacional de Energía Atómica (CNEA)
Insurgentes Sur 1079
3° piso
México, D.F.

Created in 1979 to replace the old Instituto Nacional de Energia Nuclear, which controlled the import, export, and use of all radioactive materials. The new CNEA engages in research and marketing.

Also created in 1979, along with the CNEA, were: the Instituto Nacional de Investigación Nuclear (National Institute of Nuclear Investigation); URAMEX, to explore and exploit industrial uranium; and a Safeguards and Nuclear

Security Division of the Secretariat of Natural Resources and Industrial Development.

NUCLEAR POWER REACTORS:

Under Construction:

Laguna Verde-1, at Alto Lucero, Veracruz. Boiling light-water-moderated and cooled reactor, operated by the Comisión Federal de Electricidad. Net power output: 654 MW. Commercial operation: 1982.

Laguna Verde-2, at Alto Lucero, Vercruz. Boiling light-water-moderated and cooled reactor, operated by the Comisión Federal de Electricidad. Net power output: 654 MW. Commercial operation: 1983.

## NETHERLANDS

GOVERNMENTAL AGENCIES:

Interdepartmental Committee on Nuclear Energy
c/o the Ministry of Economic Affairs
Directorate of Nuclear Energy
Bezuidenhoutseweg 6
P.O. Box 20101
2500 EC's - Gravenhage
Netherlands

Established in 1964. Prepared the policy of various ministries with respect to the peaceful use of atomic energy. Members are representatives of most ministries.

Industrial Council for Nuclear Energy
Bezuidenhoutseweg 6
The Hague
Netherlands

Advises ministers of state on industrial applications of nuclear energy.

SEMI-GOVERNMENTAL AGENCY:

Wetenschappelijke Raad voor de Kernenergie (Science Council for Nuclear Affairs)
Thorbeckelaan 360
The Hague
Netherlands

Founded 1962. Advises the state and private institutions on nuclear research.

NONGOVERNMENTAL ORGANIZATIONS:

Aktie Strohalm
Oudegracht 42
3511 AR Utrecht
Netherlands 030 314 314

Concerned for a wide range of environmental problems, including nuclear energy. Calls for a halt to current growth patterns and a reassessment of the means and ends of human life. Pub: Booklets, pamphlets, posters.

Landelijk Energie Komitee (National Energy Committee)
Tweede Weteringplantsoen 9
Amsterdam
Netherlands 020-221369

Cooperative of political parties, environmental groups and religious organizations, including Aktie Strohalm (which see), Gezamenlijke Energiekomitees Zuid-Nederland, NIVON, PPR, PSP, Stroomgroep Stop Kernenergie/Kalkar, Verbond van Wetenschappelijke Onderzoekers (VWO), and de Vereniging Milieudefense (VMD). Works for an energy policy based on three main principles: ecologically sustainable energy resources (phase out fossil fuels, get rid of nuclear power, use water power); energy conservation; and democratic supply of energy and of information about energy. The Komitee is active in the anti-nuclear movement, and is interested in the relation between nuclear power and atomic weapons, especially in Third World countries. Pub: Journal (a record of international, national, local, and regional events, and discussion of anti-nuclear strategies), bi-m.

Netherlands Energy Research Foundation -- ECN
112 Scheveningseweg
The Hague
Netherlands

Formerly: Reactor Centrum Nederland RCN. Name changed in 1976, at which time the organization's scope broadened from research on nuclear energy to research concerned with the whole field of energy supply.

World Information Service on Energy (WISE)
2e Weteringplantsoen 9
1017 ZD Amsterdam
Netherlands 20-255064

Nonprofit membership organization founded in Amsterdam in 1978, "to facilitate communication between people actively involved in the struggle for a non-nuclear world. to help the movement become more effective in countering the transnational, centralized and undemocratic forces behind nuclear 'options.'" WISE functions as an information switchboard, receiving information from various groups around the world, translating it into various languages, and publishing it in bulletins and news communiques. It also handles individual requests for information from people around the world, including members of the press. It participates in special projects, such as the "Keep It In the Ground" projects of uranium miners and native peoples, and in the organization of an upcoming international anti-nuclear strategy conference. WISE holds an annual meeting. Its activities are funded in part by the Foundation for Transnational Energy Information (the Smiling Sun Foundation). Pub: WISE Bulletin (English), bi-m.; Bulletin in Italian; other releases in various languages.

WISE contact offices are listed on the following page.

+Service Mondial d'Information sur l'Energie
13 rue Hobbema
1040 Brussels
Belgium 2-7330415

+Servicio Mundial de Informacion sobre la Energia
Calle Bruc 26º
Barcelona 10
Spain 303014939

+Servizio Mondial d'Informazione Energetica
Via Fillippini 25a
37121 Verona
Italy 45-49197

+Weltweiter Energie Informationsdienst
Arndtstrasse 17
1 Berlin 61
West Germany 30-6925637

+WISE--Denmark
c/o OOA
Skindergade 26
1159 Copenhagen
Denmark 1-110973

+WISE--Finland
c/o EVY
Valpurintie 6
00270 Helsinki 27
Finland 1-417724

+WISE--United Kingdom
34 Cowley Rd.
Oxford
United Kingdom 865-725354

+WISE--USA
1536 Sixteenth St., N.W.
Washington, D.C. 20036 202-387-0818

NUCLEAR POWER REACTORS:

Operating:

Borssele, at Borssele, Zeeland. Pressurized light-water-moderated and cooled reactor, operated by Provinciale Zeeuwse Energie Maatschappij. Net power output: 447 MW. Commercial operation since 1973.

Dodewaard, at Dodewaard, Gelderland. Boiling light-water-moderated and cooled reactor, operated by Kernkraftwerk Neckar. Net power output: 51.5 MW. Commercial operation since 1969.

## NEW ZEALAND

New Zealand's Power Planning Committee's 1977 report deleted the introduction of energy from nuclear power stations from its 15 year plans. The Royal Commission on Nuclear Power Generation, established September 1976, completed hearings and reported to the Governor-General in 1978. The Commission concluded that nuclear power could, under economic considerations, play a significant role in New Zealand's energy program early in the next century, but a firm decision need not be made until 1992 or 1996. This action, in effect, tabled nuclear development in the country.

GOVERNMENTAL AGENCY:

New Zealand Atomic Energy Committee
c/o D.S.I.R.
Private Bag
Lower Hutt
New Zealand

Responsible for advising the Ministry of Science on the development of peaceful uses of nuclear energy.

NONGOVERNMENTAL ORGANIZATIONS:

New Zealand's main anti-nuclear group, the Campaign for Non-Nuclear Futures, disbanded following the Royal Commission on Nuclear Power Generation's report in 1978, its goal of keeping nuclear development out of the country seen as achieved. Upon disbandment, the Campaign turned its remaining objective -- that of working to create a decentralized, sustainable energy future -- over to ECO, the Environmental and Conservation Organisation of New Zealand, Inc.

## NIGERIA

GOVERNMENTAL AGENCY:

A Nigerian Atomic Energy Commission was created in 1976, to develop nuclear power generation in the country, to perform research on energy use, and to prospect for radioactive minerals. Nigeria has one 600 MW nuclear power project.

## NORWAY

Norway has a research reactor at Halden which is used, under OECD auspices, by several European countries, Japan, and the U.S. (see below).

In 1978, The Norwegian Royal Commission endorsed future construction of nuclear power stations, subject to safety measures for waste processing and handling. To date, there are no nuclear plants in Norway.

SEMI-GOVERNMENTAL AGENCY:

Institutt for Energieteknikk (Institute for Energy Technology)
Head Office and Research Establishment
P.O. Box 40
N-2007 Kjeller
Norway 02-712560

Formerly: Institutt for Atomenergi. An independent foundation operated according to government mandate, which was recently revised to include: performance of research, development and analyses in the field of energy, including nuclear research and other types of research for which the Institute is competent. The Institute's prior focus on nuclear power has been reduced, and it is expanding into other fields including: process control, chemical separation processes, isotope production and application. Nuclear power technology work will be maintained at much the same level as before. Operation of the Halden Reactor

Project of the OECD will continue as established by the present international agreement.

+OECD Halden Reactor Project
P.O. Box 173
N-1751 Halden
Norway

PAKISTAN

Pakistan initiated a civilian nuclear power program ("Atoms for Peace") in the 1950s. In subsequent years, it purchased nuclear materials and technology from the U.S. and Canada, and received training from the U.S., West Germany, Canada, the U.S.S.R., and the IAEA. By 1972, Pakistan claimed to have more than 550 qualified nuclear scientists and engineers. In 1976, France agreed to sell Pakistan a nuclear reprocessing facility, but later withdrew its agreement. Pakistan now is reported to be building its own small enrichment plant at Kahota, near Rawalpindi, to obtain highly-enriched uranium; helping in the operation, it is believed, is a Pakistani scientist who worked for some time at the Dutch enrichment plant Almelo. Pakistan's interest in reprocessing is taken to indicate a desire to develop nuclear weapons.

Pakistan is not a party to the Nuclear Non-Proliferation Treaty of 1968, or to the Partial Test Ban Treaty of 1963. It has offered to open its nuclear operations to international inspection if India will do the same.

GOVERNMENTAL AGENCIES:

Pakistan Atomic Energy Commission (PAEC)
P.O. Box 1113
Islamabad
Pakistan

Responsible for: harnessing nuclear power for economic development and nuclear technological development as part of Pakistan's nuclear power program; establishing research centers; promoting peaceful uses of nuclear energy in agriculture, medicine, industry, hydrology; searching for indigenous nuclear mineral deposits; and training engineers, scientists and technicians for nuclear projects. Pub: Annual Report.

Pakistan Institute of Nuclear Science and Technology (PINSTECH)
Nilore
Rawalpindi
Pakistan

Founded in 1961 by the Pakistan Atomic Energy Commission. Controlled by the PAEC.

NUCLEAR POWER REACTORS:

Operating:

Kanupp, at Paradise Point, Sind. Pressurized heavy-water-moderated and cooled reactor, operated by the PAEC. Net power output: 125 MW. Commercial operation since 1972.

Planned:

Chasnupp, at Kundian, Punjab. Operated by the PAEC. Net power output: 600 MW. Commercial operation: 1986.

PARAGUAY

Paraguay has no nuclear laboratories or installations.

GOVERNMENTAL AGENCY:

Comisión Nacional de Energia Atómica (National Atomic Energy Commission)
Ministerio de Relaciones Exteriores
Asunción
Paraguay

Established 1960.

PERU

GOVERNMENTAL AGENCIES:

Instituto Peruano de Energia Nuclear (IPEN) (Peruvian Institute of Nuclear Energy)
Avda. Luis Aldana 120
Urbanización Santa Catalina
La Victoria
Apdo. 1687
Lima
Peru

Responsible for setting up extraction facilities for the extraction of uranium from copper and lead mines in the Andes. Planned to set up a nuclear research center by 1980.

Centro Superior de Estudios Nucleares (CSEN) (High Center for Nuclear Studies)
Lima
Peru

Research activites, in cooperation with Peruvian universities.

PHILIPPINES

GOVERNMENTAL AGENCY:

Philippine Atomic Energy Commission
Don Mariano Marcos Ave.
Diliman
Quezon City, D-505
Philippines

Established in 1958 as the official body for nuclear energy matters in the Islands. Operates under the authority of the Philippines Department of Energy. Conducts research, provides technical services to research agencies and educational institutions. Receives technical assistance from the IAEA, the Agency for International Development, and countries with which it has entered into bilateral agreements.

NUCLEAR POWER REACTORS:

Under construction:

PNPP-1, at Morong, Bataan. Pressurized light-water-moderated and cooled reactor, operated by the Philippine National Power Company. Net power output: 620.8 MW. Commercial operation: 1982.

## POLAND

Poland is a member of the Joint Institute for Nuclear Research in Dubna, near Moscow.

NUCLEAR POWER REACTORS:

Planned:

Zarnowiec-1, at Zarnowiec, Gdansk. Pressurized light-water-moderated and cooled reactor, operated by North of Poland Power Authority. Net power output: 408 MW. Commercial operation: 1985.

Zarnowiec-2, at Zarnowiec, Gdansk. Pressurized light-water-moderated and cooled reactor, operated by North of Poland Power Authority. Net power output: 408 MW. Commercial operation: 1986.

## ROMANIA

Romania is a member of the Joint Institute for Nuclear Research in Dubna, near Moscow.

GOVERNMENTAL AGENCY:

State Committee for Nuclear Energy
P.O. Box 5203
Măgurele
Bucharest
Romania

NUCLEAR POWER REACTORS:

Planned:

Olt. Pressurized light-water-moderated and cooled reactor, operated by the state. Net power output: 408 MW. Commercial operation: 1983.

## SOUTH AFRICA

South Africa is pursuing nuclear power as an economical source of energy. In its early years of nuclear development, it was aided by the U.S. and France. Recently, however, international assistance and nuclear sales have been hard for South Africa to get. It was expelled from the International Atomic Energy Agency in June of 1977. Since 1978, the U.S. has embargoed uranium shipments to South Africa.

It is believed that South Africa exploded a nuclear device on September 22, 1979, on which date a nuclear explosion was registered in the Southern Hemisphere. South Africa has denied responsibility for it, however. South Africa does have a nuclear test installation in the Kalahari Desert.

GOVERNMENTAL AGENCIES:

Atomic Energy Board (AEB)
Private Bag X256
Pretoria 0001
Transvaal
South Africa

Twelve-member statutory body of the Department of Mines. Controls radioactive materials, licenses nuclear facilities. Performs research into nuclear power economics and reactor systems. The AEB is working with South Africa's Electricity Commission (ESCOM) and other government departments to determine the feasibility of nuclear power development for South Africa. Pub: Annual Report; Nuclear Active, biennial; Pel Report, irr.

National Institute for Metallurgy (NIM)
Private Bag X3015
2125 Randburg
Transvaal
South Africa

Originally the Minerals Research Laboratory. NIM is the Extraction Metallurgy Division of the Atomic Energy Board. It is responsible for work on nuclear raw materials processing. Pub: Annual Report.

National Nuclear Research Centre
Pelindaba
Private Bag X256
Pretoria 0001
Transvaal
South Africa

Nuclear research. Home of the SAFARI-I reactor.

Uranium Enrichment Corporation of South Africa (UCOR)

A statutory body of the Department of Mines, created April 1971. UCOR has completed a pilot enrichment plant.

NUCLEAR POWER REACTORS:

Under construction:

Duynefontein, Western Cape Province.

Koeberg. Pressurized water reactors (2). Net power output: 921 MW each. Commercial operation: 1982 or 1983.

## SPAIN

GOVERNMENTAL AGENCY:

Junta de Energia Nuclear (Atomic Energy Board)
Avenida Complutense 22
Ciudad Universitaria
Madrid 3
Spain

Large organization: more than 2,000 members. Pub: Energia Nuclear, bi-m.

NONGOVERNMENTAL ORGANIZATION:

Comité Antinuclear de Catalunya
Bruc, 26, 2.
Barcelona 10
Spain 93-3015248

Works "to inform the inhabitants of Catalonia about the dangers and negative implications of nuclear power." Organizes activities opposing nuclear plants and uranium mines planned in the area. Promotes alternative energy programs. Pub: BIEN (Boletin Informative sobre Energia Nuclear), bi-m.

NUCLEAR POWER REACTORS:

Spain's National Energy Plan calls for 10,500 MW of installed nuclear power capacity by 1987.

Operating:

Jose Cabera-1, at Almonacid de Zorita, Guadalajara. Pressurized light-water-moderated and cooled reactor, operated by Union Electrica SA. Net power output: 153.2 MW. Commercial operation since 1969.

Santa Maria de Garona, at Santa Maria de Garona, Burgos. Boiling light-water-moderated and cooled reactor, operated by Centrales Nucleares del Norte SA. Net power output: 440 MW. Commercial operation since 1971.

Vandellos-1, at Vandellos, Tarragona. Gas-cooled, graphite-moderated reactor, operated by the Sociedad Hispano-Francesa de Energia Nuclear SA. Net power output: 480 MW. Commercial operation since 1972.

Under construction:

Almaraz-1 and -2, at Almaraz, Caceres. Pressurized light-water-moderated and cooled reactors, operated by Compania Sevillana de Electricidad, Hidroelectrica Espanola SA, and Union Electrica SA. Net power output: 900 MW each. Commercial operation: 1980.

Asco-1 and -2, at Asco, Tarragona. Pressurized light-water-moderated and cooled reactors, operated by Fuerzas Electricas de Cataluna SA. Net power output: 881.5 MW each. Commercial operation: 1980.

Cofrentes, at Cofrentes, Valencia. Boiling light-water-moderated and cooled reactor, operated by Hidroelectrica Espanola SA. Net power output: 935 MW. Commercial operation: 1981.

Lemoniz-1, at Lemoniz, Vizcaya. Pressurized light-water-moderated and cooled reactor, operated by Iberduero SA. Net power output: 902 MW. Commercial operation: 1981.

Lemoniz-2, at Lemoniz, Vizcaya. Pressurized light-water-moderated and cooled reactor, operated by Iberduero SA. Net power output: 902 MW. Commercial operation: 1983.

Planned:

Regodola, at Jove Lugo. Pressurized light-water-moderated and cooled reactor, operated by Fuerzas Electricas de Cataluna SA. Net Power output: 900 MW. Commercial operation: 1986.

Sayago, at Moral de Sayago, Zamora. Pressurized light-water-moderated and cooled reactor, operated by Iberduero SA. Net power output: 1036 MW. Commercial operation: 1984.

Trillo-1, at Trillo, Guadalajara. Pressurized light-water-moderated and cooled reactor, operated by Union Electrica SA. Net power output: 990 MW. Commercial operation: 1987.

Trillo-2, at Trillo, Guadalajara. Pressurized light-water-moderated and cooled reactor, operated by Union Electrica SA. Net power output: 1000 MW. Commercial operation: 1988.

Valde-Caballeros-1, at Badajoz, Valde-Caballeros. Boiling light-water-moderated and cooled reactor, operated by Hidroelectrica Espanola SA and Compania Sevillana de Electricidad. Net power output: 938.8 MW. Commercial operation: 1984.

Valde-Caballeros-2, at Badajoz, Valde-Caballeros. Boiling light-water-moderated and cooled reactor, operated by Hidroelectrica Espanola SA and Compania Sevillana de Electricidad. Net power output: 938.8 MW. Commercial operation: 1985.

Vandellos -2, at Vandellos, Tarragona. Pressurized light-water-moderated and cooled reactor, operated by ANV. Net power output: 930 MW. Commercial operation: 1984.

Vandellos -3, at Vandellos, Tarragona. Pressurized light-water-moderated and cooled reactor, operated by Fuerzas Electricas de Catàluna SA. Net power output: 1000 MW. Commercial operation: 1987.

## SWEDEN

In March 1980, the Swedes voted on a nuclear power referendum to determine the future of nuclear development in the country. Of three options -- to halt development entirely, to continue it full force, or to complete the present twelve reactor plan and then halt development -- the voters favored the latter with 58 per cent of the vote. This option was favored by the Conservative political faction, which is generally pro-nuclear, and the Liberal and Social Democrats, who have called for a period of time in which to develop renewable energy sources. Opposing nuclear development entirely are the Center and Communist parties.

GOVERNMENTAL AGENCIES:

Statens Kaernkraftinspection (Swedish Nuclear Power Inspectorate)
Box 27 106
102 52 Stockholm
Sweden

Statens Kaernkraftinspection (cont.):

Controls and inspects nuclear installations and nuclear materials.

Naturvetenskapliga forskningsradet (The Natural Science Research Council)
Box 23136
104 35 Stockholm
Sweden

Created 1977, consolidating the Swedish Atomic Research Council, and an already existing Natural Science Research Council. Administrative board under the authority of the Ministry of Education. Allocates government funds for basic research in many natural science e areas, including nuclear power. Advises the Swedish government and authorities on nuclear research. Represents Sweden in the European Organization for Nuclear Research (CERN) and NORDITA (Nordic Institute for Theoretical Atomic Physics).

AB Atomenergi (The Swedish Atomic Energy Company)
Studsvik
Fack 611 01
Nykoeping
Sweden

State-owned development corporation and national laboratory for applied research in nuclear and other energy fields. Staff of 1,000. Plans for a full-scale uranium milling operation are being studied at a pilot mill in Ranstad.

NONGOVERNMENTAL ORGANIZATIONS:

Miljoefoerbundet
Box 51
751 03 Uppsala
Sweden 018-135560

Independent organization of 80 environmental groups. Works against nuclear power and for alternative energy sources. Pub: Miljoetidningen, m.

The People's Campaign Against Nuclear Power

Umbrella organization for various anti-nuclear groups, organized prior to the Swedish referendum in March 1980. The organizational structure of the Campaign after the referendum has not yet been determined.

NUCLEAR POWER REACTORS:

Operating:

Barsebaeck-1, at Barsebaeck, Malmo. Boiling light-water-moderated and cooled reactor, operated by Sydsvenska Kraftaktiebolaget. Net power output: 570 MW. Commercial operation since 1975.

Barsebaeck-2, at Barsebaeck, Malmo. Boiling light-water-moderated and cooled reactor, operated by Sydsvenska Kraftaktiebolaget. Net power output: 570 MW. Commercial operation since 1977.

Oskarshamn-1, at Oskarshamn, Kalmar Lan. Boiling light-water-moderated and cooled reactor, operated by Oskarshamnsverkets Kraftgrupp AB. Net power output: 440 MW. Commercial operation since 1972.

Oskarshamn-2, at Oskarshamn, Kalmar Lan. Boiling light-water-moderated and cooled reactor, operated by Oskarshamnsverkets Kraftgrupp AB. Net power output: 570 MW. Commercial operation since 1974.

Ringhals-1, at Ringhals, Vaeroebacka. Boiling light-water-moderated and cooled reactor, operated by the Swedish State Power Board. Net power output: 750 MW. Commercial operation since 1976.

Ringhals-2, at Ringhals, Vaeroebacka. Pressurized light-water-moderated and cooled reactor, operated by the Swedish State Power Board. Net power output: 800 MW. Commercial operation since 1975.

Under construction:

Forsmark-1, at Oesthammar, Uppsala. Boiling light-water-moderated and cooled reactor, operated by the Swedish State Power Board. Net power output: 899 MW. Commercial operation: 1979.

Forsmark-2, at Oesthammar, Uppsala. Boiling light-water-moderated and cooled reactor, operated by the Swedish State Power Board. Net power output: 899 MW. Commercial operation: 1980.

Forsmark-3, at Oesthammar, Uppsala. Boiling light-water-moderated and cooled reactor, operated by the Swedish State Power Board. Net power output: 1060 MW. Commercial operation: 1984.

Ringhals-3, at Ringhals, Vaeroebacka. Pressurized light-water-moderated and cooled reactor, operated by the Swedish State Power Board. Net power output: 912 MW. Commercial operation: 1979.

Ringhals-4, at Ringhals, Vaeroebacka. Pressurized light-water-moderated and cooled reactor, operated by the Swedish State Power Board. Net power output: 912 MW. Commercial operation: 1980.

Planned:

Barsebaeck-3, at Barsebaeck, Malmo. Boiling light-water-moderated and cooled reactor, operated by Sydsvenska Kraftaktiebolaget. Net power output: 900 MW. Commercial operation: 1987.

Oskarshamn-3, at Oskarshamn, Kalmar. Boiling light-water-moderated and cooled reactor, operated by Oskarshamnsverkets Kraftgrupp AB. Net power output: 1000 MW. Commercial operation: 1987.

## SWITZERLAND

In February 1979, a Swiss initiative to subject nuclear power and nuclear waste disposal development to local citizen approval was rejected by 51.2 per cent of those voting. Though it failed, the initiative, and the passionate campaign in its favor, led to a public law calling for submission of nuclear plant construction to parliamentary approval. The Swiss government declared, after the initiative, that only a few nuclear power stations will be built, with a focus instead on conservation and renewable energy. A federal report published in May 1979, however, outlined a fairly large nuclear program for the next twenty years.

Switzerland has cooperated in nuclear matters with France, Brazil, West Germany, Belgium, Romania, and Canada.

GOVERNMENTAL AGENCIES:

Schweizerischen Nationalfonds zur Foerderung der Wissenschaftliscen Forschung (Swiss National Science Foundation)
Wildhainweg 20
3001 Bern
Switzerland

Nuclear research.

Eidgenoessisches Institut fuer Reaktorforschung (Swiss Federal Institute for Reactor Research)
Wuerenlingen
Switzerland

NONGOVERNMENTAL ORGANIZATIONS:

Internationale Koordinationskonferenz der Anti-Atomenergie-Bewegung

See: International section. Coordinates national as well as international efforts.

Nationale Koordination der Schweizer Atomkraftwerk-Gegner-Organisationen (National Coordination of the Swiss Anti-Nuclear Organizations)
Postfach 2409
3001 Bern (Postcheck 30-792)
Switzerland 0041-1-242 7 29

Popular initiative against a national nuclear energy program. Demands a new referendum to be held in the next four to five years on the issue of nuclear energy. Concerned with nuclear waste disposal and with exports of nuclear technology to Argentina. Pub: Leaflets, periodicals.

Schweizerische Energiestiftung (SES) (Swiss Energy Foundation)
Auf der Mauer 6
CH-8001 Zuerich
Switzerland 01-69 13 23

Works for a stabilization of energy needs, decentralization of energy supply, and the adoption of alternative energy technologies. Opposed to nuclear energy on the basis that it leads to undemocratic social structures and that radioactive waste poses an unacceptable danger. Pub: SES-Notizen, q.; Geological Aspects of Radioactive Fallout in Switzerland (in German), book; other publications, list available.

## TURKEY

GOVERNMENTAL AGENCY:

Turkish Atomic Energy Commission
Prime Minister's Office
Bestekar Sokak 29
Ankara
Turkey

Controls the development of peaceful uses of nuclear energy. Twelve members.

NUCLEAR POWER REACTORS:

Planned:

Akkuyu, at Akkuyu, Mersin. Operated by the Turkish Electric Authority. Net power output: 620 MW. Commercial operation: 1985.

## UNITED KINGDOM

The United Kingdom has extensive investments in nuclear power. It has numerous nuclear power plants (see below) which generate upwards of 15 per cent of the country's electricity. The U.K. has a fast breeder reactor program. It maintains fuel reprocessing facilities on a commercial scale at the Windscale Works.

The U.K. has nuclear weapons capacity, and has exploded both fission and fusion devices.

GOVERNMENTAL AGENCIES:

United Kingdom Atomic Energy Authority (UKAEA)
11 Charles II St.
London SW1Y 4QP
United Kingdom

A public corporation, similar to a private industry. The Authority is mandated to oversee nuclear affairs in general for the government; to research reactor systems, environmental impacts of nuclear power, and safety factors; to research and develop reactors and technology for Britain's nuclear industry; and to apply its expertise in areas other than nuclear power.

The safety regulation division of the UKAEA is the Safety and Reliability Directorate:

+Safety and Reliability Directorate
Wigshaw Lane
Culcheth, Warrington
Cheshire WA3 4NE
United Kingdom

UKAEA maintains several nuclear laboratories, including:

+Atomic Energy Research Establishment
Harwell, Didcot
Oxfordshire OX11 oRA
United Kingdom

+Culham Laboratory
Abingdon
Oxfordshire OX14 3DB
United Kingdom

United Kingdom

+Dounreay Nuclear Development Establishment
Thurso
Caithness KW14 7TZ
United Kingdom

+Risley Nuclear Power Development Laboratories
Risley, Warrington
Cheshire WA3 6AT
United Kingdom

+Springfields Nuclear Power Development Laboratories
Springfields, Salwick
Preston, Lancs. PR4 oRR
United Kingdom

+Windscale Nuclear Power Development Laboratories
Windscale, Sellafield
Seascale, Cumbria CA20 1PF
United Kingdom

The Radiochemical Centre Ltd.
Amersham, Bucks.
United Kingdom

A part of the UKAEA until 1971, when it became a limited company. Responsible for producing and marketing nuclear technology throughout the world.

British Nuclear Fuels Limited (BNFL)
Risley, Warrington
Cheshire WA3 6AT
United Kingdom

Created in 1971. Received all business responsibilities of the Production Group of the UKAEA.

NONGOVERNMENTAL ORGANIZATIONS:

British Nuclear Energy Society (BNES)
The Secretariat
Institution of Civil Engineers
1-7 Great George St.
Westminster
London SW1P 3AA
United Kingdom 01-222-7722

Established in 1962 to succeed the British Nuclear Energy Conference, and "to provide a forum for discussion and presentation of papers on nuclear energy topics." The Society is composed of 11 constituent bodies whose interests determine the exact nature of its activities. They include: Institution of Civil Engineers, Institution of Mechanical Engineers, Institution of Electrical Engineers, Institution of Electronics & Radio Engineers, Institution of Chemical Engineers, Institute of Physics, The Metals Society, Royal Institute of Chemistry, Institute of Fuel, Joint Panel on Nuclear Marine Propulsion, and Institute of Measurement and Control. Generally, activities are directed toward broad aspects of nuclear energy, and include two or three annual conferences, a lecture program arranged in conjunction with the conferences and featuring British and overseas speakers, and joint and co-sponsored events and meetings with organizations. Pub: Nuclear Energy: Journal of the British Nuclear Energy Society, 6 times/yr.; Radiation Protection in Nuclear Power Plants and the Fuel Cycle, proceedings of international conference held in 1978; technical reports in paperback form, complete list available.

Campaign ATOM
99, Woodstock Rd.
Oxford
United Kingdom

Organized under the auspices of the Campaign for Nuclear Disarmament (see below), to protest local siting of Cruise missiles. Supported by local trade unionists, churches and political parties.

Campaign for Nuclear Disarmament (CND)
29, Great James St.
London WC1N 3EY
United Kingdom 01-242-0362

Organizes meetings and activities to protest British government and international nuclear projects, especially nuclear weapons projects such as the Cruise missile. Calls on Britain to set an example for the world by renouncing nuclear weapons for the sake of world peace, and withdrawing its membership from NATO if need be. Pub: SANITY, magazine, 6 times/yr.

Centre for Alternative Industrial and Technological Systems (CAITS)
North East London Polytechnic
Longbridge Road
Dagenham
Essex RM8 2AS
United Kingdom

Research and publishing organization interested in employment, new technologies, alternatives to nuclear technology, labor organizations. Pub: Energy Options and Employment, book; The CAITS Handbook; CAITS Quarterly Bulletin; others, complete list available.

Colonialism and Indigenous Minorities Research and Action (CIMRA)
5 Caledonian Rd.
London
United Kingdom 01-226-3479

Radical support group for indigenous peoples' movements, especially those opposed to nuclear fuel processes carried out on their traditional or sacred lands. Coordinates activities with land rights movements in Australia and the Pacific, North and South America, and elsewhere. World resource center on uranium mining and enrichment. Pub: Natural Peoples News, q.

Counter Information Services (CIS)
9 Poland St.
London W1
United Kingdom 01-439-3764

Journalists' collective which publishes information not available in the established media. Aims "to investigate the major social and economic institutions that govern our daily lives in order that the basic facts and assumptions behind them be as widely known as possible." Pub: The Nuclear Disaster, "a damning analysis of Britain's nuclear industry and the companies involved"; other reports, complete list available.

Scottish Campaign to Resist the Atomic Menace (SCRAM)
2a Ainslie Place
Edinburgh 3
Scotland
United Kingdom 031-225-7752

Independent voluntary organization established in 1975. Takes its acronym from the nuclear industry term for an emergency shutdown of a nuclear reactor. Aims to inform the public of present and proposed nuclear developments and their social, political and environmental consequences; to oppose non-violently further development of nuclear power in Scotland and elsewhere; to press for a long term energy strategy based on conservation and renewable sources of energy. Organizes talks, exhibitions, film shows, peaceful demonstrations; directs efforts towards local and national government, schools and universities, and trade unions. Has been active in campaigning against the proposed Torness nuclear power station. Operates as a coordination center for other anti-nuclear groups in Scotland. Pub: SCRAM Energy Bulletin, bi-m.

Socialist Environmental and Resources Association (SERA)
9 Poland St.
London W1V 3DG
United Kingdom 01-4393749

Socialist organization established in 1973. Concerned with various aspects of the "resource/development crisis." SERA's energy group works to end further development of nuclear power and promotes in its stead conservation and coal technology. Anti-nuclear efforts are directed towards the trade union movement. Main themes of the drive are: the dangers of radioactivity exposure to workers and to the public, and the creation of more jobs which would follow from development of alternative energy technologies. Encourages citizen lobbying of elected local and national government officials. Pub: SERA News, 3 times/m.; other pamphlets.

World Energy Conference

See: International section. Works on national as well as international level.

NUCLEAR POWER REACTORS:

Operating:

Berkeley, at River Severn, Gloucestershire. Gas-cooled, graphite-moderated reactor, operated by the Central Electricity Generating Board. Net power output: 2 x 143 MW. Commercial operation since 1962.

Bradwell, at Blackwater Estuary, Essex. Gas-cooled, graphite-moderated reactor, operated by the Central Electricity Generating Board. Net power output: 2 x 125 MW. Commercial operation since: 1962.

Calder Hall, at Seascale, Cumbria. Gas-cooled, graphite-moderated reactor, operated by British Nuclear Fuels Ltd. Net power output: 4 x 50 MW. Commercial operation since 1956-1959.

Chapelcross, at Annan, Scotland. Gas-cooled, graphite-moderated reactor, operated by British Nuclear Fuels Ltd. Net power output: 4 x 49.5 MW. Commercial operation since 1959-1960.

Dungeness-A, at Dungeness Point, Kent. Gas-cooled, graphite-moderated reactor, operated by the Central Electricity Generating Board. Net power output: 2 x 205 MW. Commercial operation since 1965.

Hinkley Point-A, at Hinkley Point, Somerset. Gas-cooled, graphite-moderated reactor, operated by the Central Electricity Generating Board. Net power output: 2 x 230 MW. Commercial operation since 1965.

Hinkley Point-B, at Hinkley Point, Somerset. Advanced gas-cooled, graphite-moderated reactor, operated by the Central Energy Generating Board. Net power output: 2 x 616 MW. Commercial operation since 1977.

Hunterston-A, at West Kilbride, Scotland. Gas-cooled, graphite-moderated reactor, operated by the South of Scotland Electricity Board. Net power output: 2 x 150 MW. Commercial operation since 1964.

Hunterston-B, at Hunterston, Ayrshire. Advanced gas-cooled, graphite-moderated reactor, operated by the South of Scotland Electricity Board. Net power output: 2 x 616 MW. Commercial operation since 1977.

Oldbury-A, at Oldbury-on-Severn, Gloucestershire. Gas-cooled, graphite-moderated reactor, operated by the Central Electricity Generating Board. Net power output: 2 x 205 MW. Commercial operation since 1968.

PFR Dounreay, at Dounreay, Caithness. Fast breeder reactor, operated by the UKAEA. Net power output: 230 MW. Commercial operation since 1977.

Sizewell-A, at Sizewell, Suffolk. Gas-cooled, graphite-moderated reactor, operated by the Central Electricity Generating Board. Net power output: 2 x 210 MW. Commercial operation since 1966.

Trawsfynydd, at Merionethshire, Wales. Gas-cooled, graphite-moderated reactor, operated by the Central Electricity Generating Board. Net power output: 2 x 195 MW. Commercial operation since 1965.

Operating reactors (cont.):

Winfrith, at Winfrith Heath, Dorset. Steam-generating heavy-water reactor, operated by the UKAEA. Net power output: 92 MW. Commercial operation since 1968.

Windscale, at Windscale, Cumberland. Advanced gas-cooled, graphite-moderated reactor, operated by the UKAEA. Net power output: 32 MW. Commercial operation since 1963.

Wylfa, at Anglesey, Wales. Gas-cooled, graphite-moderated reactor, operated by the Central Electricity Generating Board. Net power output: 2 x 420. Commercial operation since 1971.

Under construction:

Dungeness-B, at Romney Marsh, Kent. Advanced gas-cooled, graphite-moderated reactor, operated by the Central Electricity Generating Board. Net power output: 2 x 607 MW. Commercial operation: 1979.

Hartlepool, at Hartlepool, Durham. Advanced gas-cooled, graphite-moderated reactor, operated by the Central Electricity Generating Board. Net power output: 2 x 625 MW. Commercial operation: 1980.

Heysham-A, at Heysham, Lancashire. Advanced gas-cooled, graphite-moderated reactor, operated by the Central Electricity Generating Board. Net power output: 2 x 625 MW. Commercial operation: 1981.

Planned:

CDFR. Fast breeder reactor. Net power output: 1250 MW. Commercial operation: 1986.

Heysham-B, at Heysham, Lancashire. Advanced gas-cooled, graphite-moderated reactor, operated by the Central Electricity Generating Board. Net power output: 2 x 611 MW. Commercial operation: 1986.

Torness, at Dunbar, East Lothian. Advanced gas-cooled, graphite-moderated reactor, operated by the South of Scotland Electricity Board. Net power output: 2 x 611. Commercial operation: 1986.

## UNION OF SOVIET SOCIALIST REPUBLICS

The U.S.S.R. has a fast breeder reactor program. It has nuclear weapons capability.

GOVERNMENTAL AGENCIES:

U.S.S.R. State Atomic Energy Committee
26 Staromonetnii pereulok
Moscow
U.S.S.R.

Controls all development of nuclear energy.

Scientific Institute of Energetics and Technology
Moscow
U.S.S.R.

Nuclear research.

Joint Institute for Nuclear Research
Head Post Office
P.O. 79
Moscow
U.S.S.R.

Nuclear research institute for the U.S.S.R. and Comecon countries.

NUCLEAR POWER REACTORS:

Operating:

APS-1 Obninsk, at Obninsk, Kaluga. Light-water-cooled, graphite-moderated reactor, operated by the Institute of Planning & Energetics. Net power output: 5 MW. Commercial operation since 1954.

Armenia-1, at Ararat Valley, Armenia. Pressurized light-water-moderated and cooled reactor. Net power output: 400 MW. Commercial since 1976.

Beloyarsk-1, at Sverdlovsk, Ural. Light-water-cooled, graphite-moderated reactor, operated by the Ministry of Power Stations of the U.S.S.R. Net power output: 102 MW. Commercial operation since 1964.

Beloyarsk-2, at Sverdlovsk, Ural. Light-water-cooled, graphite-moderated reactor, operated by the Ministry of Power Stations of the U.S.S.R. Net power output: 175 MW. Commercial operation since 1967.

Bilibin. Boiling light-water-moderated and cooled reactor. Net power output: 4 x 11 MW. Commercial operation since 1974-1976.

BN-350, at Mangyshalk, Shevchenko. Fast breeder reactor, operated by the State Committee for the Use of Atomic Energy in the U.S.S.R. Net power output: 135 MW. Commercial operation since 1973.

Bor-60, at Ulyanovsk. Fast breeder reactor. Net power output: 11 MW. Commercial operation since 1969.

Chernobyl-1 and -2, at Chernobyl, Ukraine. Light-water-cooled, graphite-moderated reactors. Net power output: 1000 MW each. Commercial operation since 1978.

Kola-1, at Murmansk. Pressurized light-water-moderated and cooled reactor. Net power output: 370 MW. Commercial operation since 1973.

Kola-2, at Murmansk. Pressurized light-water-moderated and cooled reactor. Net power output: 370 MW. Commercial operation since 1974.

Kursk-1, at Kursk. Light-water-cooled, graphite-moderated reactor. Net power output: 1000 MW. Commercial operation since 1976.

Kursk-2, at Kursk. Light-water-cooled, graphite-moderated reactor. Net power output: 1000 MW. Commercial operation since 1979.

Leningrad-1, at Leningrad. Light-water cooled, graphite-moderated reactor, Net power output: 1000 MW. Commercial operation since 1974.

Leningrad-2, at Leningrad. Light-water cooled, graphite-moderated reactor. Net power output: 1000 MW. Commercial operation since 1975.

Novo Voronezh-1, at Novo Voronezh, Voronezh. Pressurized light-water-moderated and cooled reactor, operated by the State Committee for the Use of Atomic Energy in the U.S.S.R. Net power output: 196 MW. Commercial operation since 1964.

Novo Voronezh-2, at Novo Voronezh, Voronezh. Pressurized light-water-moderated and cooled reactor, operated by the State Committee for the Use of Atomic Energy in the U.S.S.R. Net power output: 338 MW. Commercial operation since 1970.

Novo Voronezh-3, at Novo Voronezh, Voronezh. Pressurized light-water-moderated and cooled reactor, operated by the State Committee for the Use of Atomic Energy in the U.S.S.R. Net power output: 410 MW. Commercial operation since 1972.

Novo Voronezh-4, at Novo Voronezh, Voronezh. Pressurized light-water-moderated and cooled reactor, operated by the State Committee for the Use of Atomic Energy in the U.S.S.R. Net power output: 410 MW. Commercial operation since 1973.

VK-50, at Melekiss, Ulyanovsk. Boiling light-water-moderated and cooled reactor, operated by the Scientific Research Institute for Atomic Reactors. Net power output: 50 MW. Commercial operation since 1966.

Under construction:

Armenia-2, at Ararat Valley, Armenia. Pressurized light-water-moderated and cooled reactor. Net power output: 400 MW. Commercial operation: 1979.

BN-600, at Sverdlavsk, Beloyarsk. Fast breeder reactor. Net power output: 600 MW. Commercial operation: 1979.

Chernobyl-3. Light-water-cooled, graphite-moderated reactor. Net power output: 1000 MW. Commercial operation: 1980.

Ignalino-1, at Ingolianska, Lithuania. Light-water-cooled, graphite-moderated reactor. Net power output: 1500 MW. Commercial operation: 1984.

Ignalino-2, at Ingolianska, Lithuania. Light-water-cooled, graphite-moderated reactor. Net power output: 1500 MW. Commercial operation: 1985.

Kalinin-1. Pressurized light-water-moderated and cooled reactor. Net power output: 1000 MW. Commercial operation: 1981.

Kalinin-2. Pressurized light-water-moderated and cooled reactor. Net power output: 1000 MW. Commercial operation: 1982.

Kola-3. Pressurized light-water-moderated and cooled reactor. Net power output: 420 MW. Commercial operation: 1981.

Kola-4. Pressurized light-water-moderated and cooled reactor. Net power output: 420 MW. Commercial operation: 1982.

Kursk-3, at Kursk. Light-water-cooled, graphite-moderated reactor. Net power output: 1000 MW. Commercial operation: 1980.

Leningrad-3, at Leningrad, Lelingrad. Light-water-cooled, graphite-moderated reactor. Net power output: 1000 MW. Commercial operation: 1979.

Leningrad-4, at Leningrad, Lelingrad. Light-water-cooled, graphite-moderated reactor. Net power output: 1000 MW. Commercial operation: 1980.

Nikolajev-1, at Nikolajev, South Ukraine. Pressurized light-water-moderated and cooled reactor. Net power output: 1000 MW. Commercial operation: 1980.

Novo Voronezh-5, at Novo Voronezh, Voronezh. Pressurized light-water-moderated and cooled reactor. Net power output: 1100 MW. Commercial operation: 1979.

Rovno-1, at Rovno, West Ukraine. Pressurized light-water-moderated and cooled reactor. Net power output: 420 MW. Commercial operation: 1979.

Rovno-2 and -3, at Rovno, West Ukraine. Pressurized light-water-moderated and cooled reactors. Net power output: 420 MW each. Commercial operation: 1981.

Smolensk-1, at Smolensk, Byelorussia. Light-water, graphite-moderated reactor. Net power output 1000 MW. Commercial operation: 1980.

Smolensk-2, at Smolensk, Byelorussia. Light-water, graphite-moderated reactor. Net power output: 1000 MW. Commercial operation: 1981.

Planned:

Atash-1, in Crimea. Net power output: 1000 MW.

BN-1600. Fast breeder reactor. Net power output: 1600 MW.

Chernobyl-4. Light-water-cooled, graphite-moderated reactor. Net power output: 1000 MW. Commercial operation: 1982.

Kalinin-3. Pressurized light-water-moderated and cooled reactor. Net power output: 1000 MW. Commercial operation: 1987.

Kalinin-4. Pressurized light-water-moderated and cooled reactor. Net power output: 1000 MW. Commercial operation: 1988.

Kursk-4, at Kursk. Light-water-cooled, graphite-moderated reactor. Net power output: 1000 MW.

Nikolajev-2, -3 and -4. Pressurized light-water-moderated reactors. Net power output: 1000 MW each.

West Ukrainian-3 and -4. Pressurized light-water-moderated and cooled reactors. Net power output: 1000 MW each.

Zaporozhe-1, in the Ukraine. Net power output: 1000 MW.

## URUGUAY

### GOVERNMENTAL AGENCY:

Comisión Nacional de Energía Atómica (National Atomic Energy Commission)
Rincón 723
3 Montevideo
Uruguay

## VENEZUELA

### GOVERNMENTAL AGENCIES:

Consejo Nacional para el Cesarrollo de la Industria Nuclear (National Council for the Development of Nuclear Industry)
Apdo. 68233
Caracas 106
Venezuela

Founded 1975. President of the Council is Venezuela's Minister of Energy and Mines; Vice-President is the President of the State Electricity Industry.

Instituto Venezolano de Investigaciones Cientificas (IVIC) (National Institute for Scientific Investigations)
Alto de Pipe
Apdo. 1827
Caracas
Venezuela

Basic research. Has small research reactor.

## YUGOSLAVIA

### NONGOVERNMENTAL ORGANIZATIONS:

Boris Kidrić Institute of Nuclear Sciences
P.O. Box 522
Belgrade
Yugoslavia

Established in 1948. Nuclear research. Pub: Bulletin.

Institute for Geological and Mining Exploration and Investigation of Nuclear and Other Mineral Raw Materials
Rovinjska 12
Belgrade
Yugoslavia

### NUCLEAR POWER REACTORS:

Under construction:

Krsko, at Krsko. Pressurized light-water-moderated and cooled reactor, operated by Nuklearna Elektrana Krsko. Net power output: 632 MW. Commercial operation: 1980.

Vir, in Dalmatia.

Planned:

Zagreb, near Zagreb, Croatia.

# Indexes

# INDEX OF ORGANIZATIONS

This is an index to all types of organizations listed in this book, including selected programs of organizations.

INDEX - ORGANIZATIONS

## INDEX OF ACRONYMS AND INITIALISMS